Precalculus in Context
Functioning in the Real World

PWS New Directions in Mathematics Series

Precalculus in Context
Functioning in the Real World

Marsha J. Davis, Mount Holyoke College
Judy Flagg Moran, Trinity College
Mary E. Murphy, Smith College

Research Assistant: Anne L. Kaufman

PWS-KENT PUBLISHING COMPANY
Boston

PWS-KENT
PUBLISHING COMPANY

20 Park Plaza
Boston, Massachusetts 02116

©1993 PWS-KENT Publishing Company.

All rights reserved. No part of this book may be reproduced, stored in a retrieval system, or transcribed, in any form or by any means—electronic, mechanical, photocopying, recording, or otherwise—without the prior written permission of PWS-KENT Publishing Company.

PWS-KENT Publishing Company is a division of Wadsworth, Inc.

ISBN 0-534-19788-4

Cover photo of irrigation by Barrie Rokeach.

Acquisitions Editors: Anne Scanlan-Rohrer, Tim Anderson
Editorial Assistant: Leslie With
Production Editor: Carol Lombardi
Designer: Andrew Ogus
Print Buyer: Diana Spence
Compositor: Rachel Goldeen
Printer and Binder: Malloy Lithographers

Printed in the United States of America

93 94 95 96 97 — 10 9 8 7 6 5 4 3 2 1

CONTENTS

Preface   vii
Topic Comparison Guide   x
A Letter to the Student   xi
For the Student: Guidelines for Lab Reports   xiii

Laboratory 1   Fahrenheit   1
Linear functions
        Homework 1.1: The Graphing Game   7
        Homework 1.2: "Have it your way..."   11
        Homework 1.3: "A Big Moosetake"   13

Laboratory 2   Galileo   15
Quadratic functions
        Homework 2.1: "Don't Fence Me In"   21
        Homework 2.2: Exploring Quadratics   23

Laboratory 3   Graph-Trek   27
Shifting and reflecting graphs
        Homework 3.1: Vertical Stretching and Compression   35
        Homework 3.2: Horizontal Stretching and Compression   39
        Homework 3.3: Symmetry about the Y-axis   43
        Homework 3.4: Symmetry about the Origin   47
        Homework 3.5: Symmetry Test   51
        Homework 3.6: Absolute Value in Functions   53

Laboratory 4   Electric Power   59
Inequalities
        Homework 4.1: Inequalities   65
        Homework 4.2: "It's Not Easy Being Green"   69
        Homework 4.3: The Least Squares Line   71

Laboratory 5   Bordeaux   77
Multivariable linear functions
        Homework 5.1: A Boring Problem   83
        Homework 5.2: Quilts   85

Laboratory 6   Packages   89
Polynomial functions
        Homework 6.1: Exploring Polynomial Graphs: Cubics   95
        Homework 6.2: Exploring Polynomial Graphs: General   99
        Homework 6.3: Optimize for a Ban on Waste   105

Laboratory 7   **Doormats**   109
Rational functions
         Homework 7.1: Black Holes and Vertical Asymptotes   113
         Homework 7.2: Long-term Behavior of Rational Functions   115

Laboratory 8   **More Power**   119
Multivariable nonlinear functions
         Homework 8.1: "Lost in the Supermarket"   123

Laboratory 9   **AIDS**   127
Exponential growth functions
         Homework 9.1: Growth Rates of Exponential Functions   131
         Homework 9.2: Limits to Growth   133
         Homework 9.3: "Vamoose"   135

Laboratory 10   **Radioactive Decay**   139
Exponential decay functions
         Homework 10.1: Changing the Initial Amount of Substance   145
         Homework 10.2: Repeated Doubling versus Repeated Halving   147
         Homework 10.3: Exponential Decay   149

Laboratory 11   **Earthquakes**   151
Logarithmic functions
         Homework 11.1: Explorations of ln x 159
         Homework 11.2: Algebraically Equivalent Functions   161
         Homework 11.3: Alternate Formula for $M(x)$ 163
         Homework 11.4: "I Hear the Train a-Comin'..."   165

Laboratory 12   **Daylight and SAD**   167
Sine functions
         Homework 12.1: SAD and Latitude   173
         Homework 12.2: An Alternate Variety of SAD   175
         Homework 12.3: In the Bath   177

Appendix: Using the TI-81 Graphing Calculator: A Primer   181

Preface

Precalculus in Context can serve as a valuable companion to any precalculus text. The twelve projects are designed to form an integral part of a college or high-school precalculus course. Except for two sections on multivariable functions, the topics are those you'd expect to find in a one-semester course and are presented in the traditional order, beginning with linear and quadratic functions and ending with exponential, logarithmic, and trigonometric functions.

Here, however, the similarity ends. We have attempted to avoid the compartmentalized structure of traditional precalculus by weaving several themes through the text:

- mathematical modeling of real-life phenomena

- choice of a convenient and meaningful scale (including exponential and logarithmic scales)

- constant and nonconstant rates of change (and how to interpret them)

- relationships between algebraic statements and geometric representations

LABS OFFER REAL-WORLD SITUATIONS

The spirit of the book, also, is one that may be new to a student trained to think that doing mathematics means manipulating an algebraic formula, finding an "answer," and checking it in the back of the book. This book *has* no "back"! There are, however, results that are correct and others that are not, some conclusions and explanations that are better than others. There may be several reasonable ways to answer a question. Also, one-sentence or one-number answers are not appropriate for these projects. Some of the exercises encourage the students to use the graphing utility to explore relationships and discover patterns. Most of the projects plunge the students into a real-life situation and ask them to investigate the mathematics they find there. They should always be checking among themselves (rather than in the back of the book) to be sure that their results jibe with reality. Although some students may resist this approach, many others will find themselves (much to their own surprise) blossoming in a mathematics class for the first time in their lives.

What's more, the manner in which this book is intended to be used is strikingly different from that of a traditional precalculus text or workbook. Each of the twelve laboratory projects is designed for group collaboration, and each project culminates in a written report submitted by the group. Guidance is given to the students to encourage written work of high quality, in which they explain in understandable prose the mathematical ideas they've been exploring, relating them to the problem situation. The lab report is an opportunity for students to demonstrate both their grasp of the concepts and their creativity in making those concepts accessible, not simply to their mathematics instructor, but to any educated reader.

You will observe that there are no numbers on the questions posed within each laboratory project. This was deliberate. We have found that, regardless of what we may say, if there are numbered questions within a project, some students will think that the *real* goal of the lab is to find the right answers to all the numbered questions. We want, instead, to give them a format that encourages them to use the questions as guides in proceeding through the project and as hints for what they might want to include in their lab report. Their report should be a unified discussion, not a series of answers.

Each of the twelve projects is followed by associated exercises which extend the concepts of the lab, providing practice and opportunities to explore. These supplementary exercises can be done either collaboratively or independently. Some of them work particularly well as classroom exercises; others are good as homework assignments or test questions.

USE WITH ANY GRAPHIC FACILITY

The book can be used with either a graphing calculator or a computer graphing program. We have included a brief tutorial for the TI-81 graphing calculator, but we do not mean to imply that this particular calculator is more appropriate for this material than others. In fact, we teach a course at Smith College using this material in a classroom equipped with several computers and a computer-graphing package; the students do not, in general, use graphing calculators. At Mount Holyoke College, students use both a computer graphing utility and the Casio fx 7500G graphing calculator.

For most instructors, this approach constitutes a radical departure from traditional methods. We want to reassure them that the Instructor's Guide contains a set of lab-by-lab guidelines to points we look for in grading the reports, as well as hints on how to use the exercises. These are not rigid requirements, but suggestions for first-time users. Admittedly, implementing this approach requires extra effort from both instructors and students, but the rewards—excitement in the classroom, active student involvement, and opportunities to be creative—are well worth the energy expended.

Have fun!

ACKNOWLEDGMENTS

We would like to thank the New England Consortium for Undergraduate Science Education (NECUSE) for supporting our work in precalculus during the summers of 1990, 1991, and 1992. Carren Knoche (Smith College) was our energetic researcher and mathematical word processor during the summer of 1991. Bob Currier (Smith College) helped us learn to write in LATEX, and Jim Callahan (Smith College) rescued us from some technical difficulties in mathematical word processing. Jim Henle's computer graphing program, GRAPH, developed for the Five-College "Calculus in Context" project, has been an invaluable tool. Lester Senechal (Mount Holyoke College) helped us gain support for this project. Lucy Deephouse (Trinity College), Dan Carter (University of Massachusetts and Smith College), and Judith Grabiner (Pitzer College) were enthusiastic early adopters who gave us valuable feedback. We extend our thanks to the colleagues who reviewed our manuscript: Dick Clark, Portland Community College; Don Cohen, SUNY Cobleskill;

Margaret J. Greene, Florida Community College at Jacksonville; William Grimes, Central Missouri State University; and Stuart Thomas, University of Oregon. We also thank Don Gallagher and his students at Central Oregon Community College, who class-tested the text and provided feedback in the spring of 1992.

Finally, we wish to express gratitude to our students, the "test drivers" for the preliminary version of this manual.

TOPIC COMPARISON GUIDE

Laboratory Number		Ruud/Shell[1]	Holder[2]
1.	Linear functions, graphs, constant rates of change	2.1, 2.2, 2.4	3.1, 3.2, 4.3
2.	Quadratic functions, parabolas, changing rates of change	2.4	4.4
3.	Transforming a graph: shifting, reflecting, stretching, compressing, symmetry, and absolute value	2.3, 2.5, 2.6	3.3, 3.4, 4.2
4.	Inequalities	1.5	2.8, 2.9, 2.10
5.	Multivariable linear functions	—	—
6.	Polynomial functions and optimization	3.1	5.7
7.	Rational functions and long-term behavior	3.4	5.8
8.	Multivariable nonlinear functions	—	—
9.	Exponential growth functions	4.1, 4.2	6.1, 6.4
10.	Exponential decay functions	4.1	6.1, 6.4
11.	Logarithmic functions	4.3	6.2, 6.4
12.	Sine functions	5.5	7.6

---

[1] *Prelude to Calculus*, Second Edition, Warren L. Ruud, Terry L. Shell (Wadsworth Publishing Company, Belmont, CA 1993)

[2] *A Primer for Calculus*, Sixth Edition, Leonard I. Holder (Wadsworth Publishing Company, Belmont, CA 1993)

A LETTER TO THE STUDENT

Dear Mathematician,

(Yes, we mean you!)

We are inviting you to experience mathematics in a way that may be new to you, but very familiar to mathematicians and scientists.

This manual presents a series of twelve laboratory projects that you will explore with your lab partners. We expect and encourage you to be true investigators: to use experimentation, discovery, and common sense seasoned with inspired guessing to analyze the situation presented in each lab.

Like mathematicians and scientists, you will find working with your colleagues invaluable. The members of your lab group will bring distinct viewpoints and talents to the week's problem. We hope you will find that talking about mathematics with one another and then writing up your group's ideas in a lab report will clarify and refine your own understanding. In the "real world," the world we want you to investigate in these labs, mathematics is seldom a solo act.

Also, as in most of the real world, there may be no single correct answer to a laboratory exercise. Some conclusions may be formulated more clearly than others, some approximations may be more precise, but a lab report isn't "wrong" or "right" any more than is a paper you write for an economics course. Like your economics teacher, your lab instructor will be interested in how well you analyze data, how adept you are at formulating conclusions, and how coherently you express your results.

This may not feel like math to you, or, as one of our early lab students protested, "*This isn't math! You have to think about it!*" We hope that at the end of this course you will appreciate that, in fact, mathematics *does* require that you think—that you reason well and thoughtfully—but also that you wonder. One of the best ways a mathematician, or any researcher, can start a sentence is, *"What if . . . ?"*

Your colleagues,

Marsha Davis

Judy Moran

Mary Murphy

FOR THE STUDENT

Guidelines for Lab Reports

The weekly laboratory report is one of the most important components of your precalculus course. It is an excellent means for gauging how well you and the other members of your group understand the material and are able to use the concepts you're learning. Each report represents the joint work and conclusions of the group and is written by the "scribe" for that week. The office of scribe rotates among the group members, so that each student ends up writing approximately the same number of lab reports during the semester.

As soon as she writes her report, the scribe should make a copy for each member of the group. The others should review her work and make comments. She should then revise her report, if necessary, so that it reflects the consensus of the group. She should also make copies of the report for her group when she receives it back with the instructor's comments. These reports will form an important body of notes to which each student will refer during the semester and from which each will want to study at exam time.

On the last page of each lab project you will find instructions specific to that particular lab. There are, however, some general guidelines that apply to every lab.

A lab report is *not* a list of answers to the questions posed by the lab sheets. It is an *essay* showing what the group members have learned and what they now understand about the project. **Anyone should be able to pick up your report and understand what the project was about.** Think of your audience as a person who is familiar with precalculus mathematics but who has not read the lab sheets. You need to write in complete sentences and explain enough so that the reader will understand. This does not mean that the report needs to be very long; it does mean that it must be able to stand on its own.

A good lab report contains the following:

- an introduction for your reader (What was this lab about?)

- the mathematical concepts involved

- an outline of your procedure

- anything you learned along the way

- conclusions

Graphs can do much to clarify what you're saying. They needn't be elaborate, but the axes should always be labeled and the units, if appropriate, should be specified.

Be sure the pages can be photocopied. A single-sided presentation is more readable than a double-sided one.

There is room for plenty of creativity! You'll have opportunities during the semester to read other groups' reports, and you'll notice that no two are the same (nor should they be).

One tendency you'll probably have at first is to go through the lab sheets to highlight the questions you find, and then go about the business of answering those questions. Fine; but that, in itself, doesn't make a lab report. Imagine yourself as a tailor: you cut out the pieces of fabric and put them together with large basting stitches; then you sew the seams and remove the bastings. Your final suit doesn't have (we hope) any basting stitches in it. Some of the questions in the lab sheets are like basting stitches: answering them helps to shape your work and to keep you from straying too far off the track. The finished product doesn't even need to mention those questions; the fact that you produced an elegant piece of work shows that you put things together correctly.

Other questions, however, do need to be answered in the lab report. How can you tell which questions are basting stitches and which are essential construction details? Experience will help. So will the specific guidelines at the end of each project. If a lab gives you a concrete example to calculate (e.g., What is the volume when $L = 44$ in.?), that's probably just an illustration for your benefit; and the answer ($V = 7744$ cu. in.) doesn't need to be in the report unless you're using it to illustrate a concept. If, however, the lab sheet asks, Why does one shape yield more volume than the other?, your answer will form an essential part of the report and should be incorporated in your discussion.

The lab sheets help to lead you, by means of questions, from specific examples to general conclusions. Your lab report, in contrast, should begin with those conclusions and then illustrate them with whatever examples and graphs you need. In other words, although you do the project in the order suggested by the lab sheets, you might write about it more effectively by switching things around.

Laboratory 1

PREPARATION

After counting, one of the most basic uses of numbers is measurement. English-speaking countries have inherited a particularly arbitrary system of measurement. One example is the use of degrees Fahrenheit to measure temperature. The Fahrenheit scale was established in the early eighteenth century by Daniel Gabriel Fahrenheit, a German physicist. Fahrenheit obtained a reading for the uppermost fixed point on his scale by placing a thermometer under the armpit of a healthy man; he obtained the lowest, 32, by using an ice and water mixture to determine the freezing point of water. The number 0 approximated the temperature of an ice and salt mixture (which was widely considered to be the coldest possible temperature), so all readings on a Fahrenheit thermometer were assumed to be positive. Fahrenheit arbitrarily assigned to the uppermost and lowest fixed points the values 96 and 32 to eliminate "inconvenient and awkward fractions."

By contrast, the Celsius scale, like our number system, is based on the number 10. The fixed points, corresponding to the freezing and boiling temperatures of water, are labeled 0 and 100, respectively. In this country, the campaign to "convert to metric" has been waged for decades. One problem is that in order to use a system of measurement effectively, a person needs to internalize the scale. We know that a 70° Fahrenheit room temperature feels comfortable, 60° requires a sweater, and 50° a coat. But if the weather forecast is for a high of 17° Celsius, many of us would need to convert this temperature into our internalized Fahrenheit system in order to anticipate the weather correctly.

Many banks and businesses provide time and temperature signs with temperature given in both Fahrenheit and Celsius degrees to help us internalize the Celsius scale. Suppose that over the course of the semester, you collected the following data from a bank sign in your town. (Note that the sign only reports the temperature in integers!)

Degrees Celsius	Degrees Fahrenheit
−10	14
−4	25
0	32
5	41
12	54
20	68
24	75
30	86

One way to understand the relationship between the two scales is to draw a graph. On graph paper, draw a horizontal axis representing degrees Celsius and a vertical axis for degrees Fahrenheit, and plot the points corresponding to the numbers in the table. Do the points make a pattern? If so, describe it.

We call the data in the table *discrete:* each value given by the sign is separated by at least one degree from any other. However, we know that temperature varies *continuously:* for the temperature to rise from 3° to 4°, it must pass through all intermediate real values (even $\sqrt{11}$ and π!). Thus we are justified in connecting our points by a continuous curve. Do this and use your graph to estimate what temperature Celsius corresponds to 0° Fahrenheit.

Draw two more graphs of the same data: first, change the scale of the horizontal axis by doubling the distance you used on the original graph to represent one unit. (Leave the vertical axis unchanged.) Then, halve that distance. What changes about your graph and what remains the same?

Bring your graphs with you to your first lab. In Lab 1, you will be investigating the relationship between temperature measured in degrees Celsius and temperature measured in degrees Fahrenheit. You will also use your graphing utility to further explore the effects of altering the scale on one of the axes of your graph.

LABORATORY 1: FAHRENHEIT

"Some like it hot"

In your preparation for this lab, you drew three graphs of the relationship between temperature measured in degrees Celsius and temperature measured in degrees Fahrenheit. Compare your graphs with those of your lab partners to make sure you agree on how to represent the data given in the table in the Lab 1 preparation. Do you all agree on what changes and what remains the same when one of the scales is altered?

In the table, both degrees Celsius and degrees Fahrenheit vary, but the variation is not random. In fact, if we know the temperature in degrees Celsius, we can find the Fahrenheit temperature. Because the temperature measured in Fahrenheit degrees is uniquely determined, once we know the Celsius temperature, we can say that degrees Fahrenheit is a **function** of degrees Celsius. Let the letter C represent the number of degrees Celsius and F the number of degrees Fahrenheit. We call F and C **variables** (their values change with time and date). We would like to write the relationship between F and C explicitly using algebra. For F to be a function of C, this relationship would have the form

$$F = (\text{some algebraic expression involving } C).$$

An equation of this form shows how the value of F depends on the value of C, so we call F the **dependent variable** and C the **independent variable**. We want to find an equation of this form to serve as a **model** for the data in our table. (The equation is an idealization of the data; the numbers in the table may not fit the equation exactly.) That is, we want to perform the same algebraic operation on each number in the first column and obtain (approximately at least) the corresponding number in the second column.

A very simple relationship between F and C would be one in which, given a value for C, we obtain the corresponding value for F by multiplying by a fixed constant. If k represents this constant, then the equation we're looking for would be of the form $F = k \cdot C$. When the relationship between F and C is of this form, we say F is **proportional** to C.

Here's an everyday example: the amount of sales tax is proportional to the price of an item. If the state imposes a sales tax of 7% and the item costs \$22.65, the amount of sales tax is .07 times \$22.65, or \$1.59. In general, if the item costs D dollars, the amount of the sales tax is $.07D$.

Do you think F is proportional to C? Using values from the table, construct an argument to support your answer.

Now, try to construct an equation that models the relationship between F and C. (**Hint:** the third row of your table is very helpful.)

What does your model predict for the value of C corresponding to $F = 0$? Does this answer agree with your graph? Explain.

The **domain** of a function is the set of allowable values for the independent variable. We will sometimes call the independent variable the *input* variable, since if we were using a calculator to compute F, we would input -10, for example, to obtain the value 14 for F. The domain of the function defined by the table consists of eight numbers. (What are they?)

The equation modeling the data in the table can be used to compute F given *any* value of C, so the domain of the function defined by the *equation* is **R**, the set of all real numbers.

Use your graphing utility to graph the algebraic relationship between F and C that you found above. Adjust the scale on the horizontal axis so the graph looks like the first one you drew in preparation for this lab. Use your graphing utility to answer the following questions. (You may need to change the "viewing window" to observe different parts of the graph.)

Fahrenheit believed that 0° F was the coldest possible temperature. (Those of us living in New England know better!)

What value for C gives a value of approximately $F = -10$?

Is F always larger than C?

As C increases, what happens to F?

As C increases, does F change more slowly than C or more rapidly?

As C decreases, what happens to F?

Is there any temperature for which $F = C$? If so, what is this common value?

Explain how you used your graphing utility to determine your answers to these questions.

When $C = 0$, $F = 32$. When $C = 5$, $F = 41$.

From the graph, estimate F when $C = 10$.

How much has F changed from the value it had when C was 5?

How much do you think F will change between $C = 10$ and $C = 15$?

Does your graph confirm this answer?

Complete the following: for this function, a change of _____ in the value of C produces a change of _____ in the value of F. If the Celsius temperature goes up one degree, what happens to the Fahrenheit temperature?

Now look at the algebraic relationship between F and C that you postulated. Do you see a way of "reading" the information in the previous two questions from the equation involving F and C?

From your model, you see that some integer values of C produce fractional values for F and vice versa. The bank sign, however, displays only integer values for temperature. There are two conventions for doing this: rounding off, which replaces a rational number with the closest integer; and truncation, which drops the fractional part of a number. Which method do you think the bank sign uses? Use specific entries from the table to justify your answer.

THE LABORATORY REPORT

In your laboratory report, give the algebraic relationship you determined to model the relationship between F and C. Show how you found this relationship. Discuss whether or not F is proportional to C, using data from the table to justify your answer. Include a careful sketch of the graph of the function you determined and explain how the constant(s) that appear in the algebraic formula affect the graph of the function. Then present a second graph of the same function with a different scale for the horizontal axis. Explain which features of the graph are altered by a change of scale and which features remain the same. How would you explain to a friend the difference between the function given by the table of values and that given by the algebraic formula? Do you think the bank sign uses rounding off or truncation in determining its values? Explain your answer by referring to specific values in the table.

Homework 1.1: The Graphing Game

Adjust your viewing window so that the *x*- and *y*-axes have the same scale, that is, so that the distance between 0 and 1 on the *x*-axis is the same as the distance between 0 and 1 on the *y*-axis. Remember, parallel lines never intersect, and perpendicular lines meet at right angles.

1. Overlay the graphs of the functions $f(x) = x$, $g(x) = x + 1$, and $h(x) = x + 2$. Describe in words the pattern that you see. Sketch the graphs of the three functions on a single set of axes.

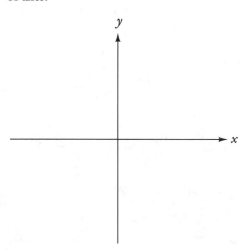

2. Overlay the graphs of the functions $f(x) = x$, $g(x) = 2x$, and $h(x) = 4x$. Sketch the graphs of the three functions on a single set of axes.

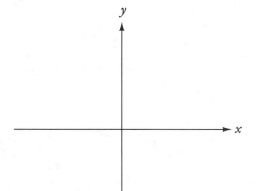

Describe in words the pattern that you see. Which of the three graphs rises most steeply? Which least steeply?

3. Overlay the graphs of the functions $f(x) = x + 1$, $g(x) = 2x + 1$, and $h(x) = 4x + 1$. How does this set of graphs compare to the set in the previous problem?

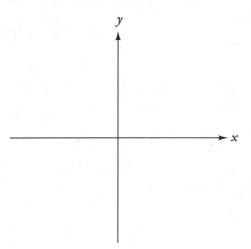

4. Overlay the graphs of $f(x) = x$ and $g(x) = -x$. Sketch what you see. Do these lines appear to be perpendicular?

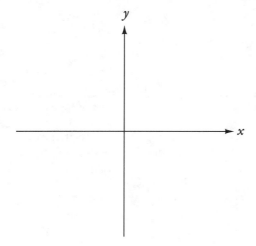

5. Graph $f(x) = -2x$. Would a line perpendicular to $f(x)$ have positive or negative slope? Overlay the graph of $g(x) = 2x$. (Be sure that the x- and y-axes still have the same scales.) Would a line perpendicular to the graph of $f(x)$ have larger or smaller slope than the line $y = 2x$?

6. Experiment with your graphing utility (using a trial-and-error approach) and find a line that appears to be perpendicular to the graph of $f(x) = -2x$. Write down the equation of this line.

Homework 1.2: "Have it your way..."

In Lab 1, you wrote an equation describing F as a function of C; that is, F was the dependent and C the independent variable. However, each Fahrenheit temperature also determines a unique Celsius temperature, so we can also write C as a function of F.

1. Write the equation that gives degrees Celsius in terms of degrees Fahrenheit and show how you determined it.

2. One group of students, working with the same table of values you were given in the Lab 1 preparation, came up with the following function to relate Fahrenheit temperatures to Celsius temperatures.

$$F = \frac{20}{11}C + 32$$

Using your graphing utility, graph this function simultaneously with the function you determined in Lab 1.

3. By substituting all the values from the table given in the preparation for Lab 1, determine how well this second function models the data. Do your results suggest any conclusions about models of real-life situations?

Homework 1.3: "A Big Moosetake"

LINEAR FUNCTIONS AND AVERAGE RATES OF CHANGE

The moose populations in Vermont, New Hampshire, and Maine have increased sharply in the past decade. Meeting a moose in the wilds can be dangerous, but meeting one head-on in a car can be fatal. Even though signs warning of "Moose Crossing" have been posted, moose-car collisions on northern New England roads are posing an increasing hazard for motorists (and the moose). The table below provides data on moose-car collisions that have occurred from 1980 to 1990[1].

	1980	1984	1985	1988	1990
Vermont	0	unknown[2]	12	unknown	41
New Hampshire	unknown	unknown	49	117	170
Maine	156	215	unknown	unknown	500

1. Consider first the data for Vermont. On a separate sheet of paper, plot the number of collisions versus the year. ("Number of collisions" will appear on the vertical axis; "year" will appear on the horizontal axis.) How many *more* collisions were there in 1990 than there were in 1985? This number represents the increase in the number of collisions over that entire five-year period. Translate this number into the **average rate of change** in the number of collisions *per year* for that period.

    Now calculate the average rate of change in the number of collisions per year for the period 1980–1985.

2. Compare the average rate of change (collisions per year) for 1980–1985 and for 1985–1990. Did the rate increase, decrease, or remain the same for the two consecutive five-year periods?

---

[1]"Comeback for Moose in New England Leads to Road Hazard," *The New York Times*, June 3, 1991.

[2]The missing numbers are not available; adequate records may not have been kept for those years. Life is like that, and we have to deal with the information we've got.

3. Write down the equation of the line that passes through the data points corresponding to the years 1980 and 1985. (Specify which year is to be "year zero" on your graph.) Sketch this line on your plot of the data. If this line had accurately represented the relationship between number of collisions and year, how many collisions would you have expected in 1990? (What would then have been the average increase in numbers of collisions per year from 1985 to 1990?)

4. Make another graph using the data for New Hampshire. Do the points lie on a line? Check by computing the average increase in numbers of collisions per year for 1985 to 1988 and then for 1988 to 1990. Are these rates constant?

5. Finally, plot the data for Maine. Draw a line that passes through the data points corresponding to the years 1980 and 1984. Draw a second line that passes through the data points corresponding to the years 1984 and 1990. Without doing any calculations, state what happens to the average rate of change in numbers of collisions. As time passes, does it increase, decrease, or remain constant? Explain how you can answer this question by simply looking at your graphs.

Laboratory 2

PREPARATION

In the first lab, we deduced a relationship between the temperature measured in degrees Fahrenheit, denoted F, and the temperature measured in degrees Celsius, denoted C. The relationship is given by the equation

$$F = \frac{9}{5}C + 32.$$

We called C the **independent variable** and F the **dependent variable** in the equation. Because we can use the equation to produce exactly one value of F for each value of the variable C, the equation is said to define F as a **function** of C.

Another famous function that models the physical world was discovered by Galileo. Until Galileo's experiments with falling bodies, many scientists accepted Aristotle's contention that heavy objects fall faster than light ones do. Galileo studied the motion by measuring the distance an object falls in a fixed amount of time. Translating his data to our own units of measure, we have the following table of values for his two variables: t, which gives the time, measured in seconds, that the object has been falling; and s, which gives the total distance the object falls, measured in feet.

t	s
0	0
1	16
2	64
3	144
4	256

Before conducting his experiments, Galileo postulated that the distance fallen was proportional to the time elapsed. Remember, in Lab 1 you learned that s is **proportional** to t if $s = k \cdot t$, where k is some fixed constant (called the constant of proportionality). Do you think Galileo's data confirmed his hypothesis? Test your answer using the data presented in the table.

Use a piece of graph paper to plot the points corresponding to the values in the table. Since it seems natural to think of s, the distance fallen, as dependent on t, the time the object has been falling, let s be the dependent variable measured on the vertical axis, and let t be the independent variable measured on the horizontal axis. Experiment with scales on the two axes to best exhibit your data.

Be sure to label both axes carefully! For *any* graph, proper labeling means specifying the following next to each axis:

- the quantity being measured (e.g., time, area, population)

- the units in which that quantity is measured (days, square meters, thousands of persons, etc.)

15

You also need to show a scale on each axis. The numbers in the scale should be equally spaced; that is, the length of axis from $y = 0$ to $y = 10$, for instance, should be the same as the length from $y = 40$ to $y = 50$.

Bring your graph(s) with you to the lab. In Lab 2, you will follow in Galileo's footsteps by finding a function relating s and t that models the data in the table. You will also use your graphing utility to investigate how a change in the independent variable *time* produces a change in the dependent variable *distance*.

LABORATORY 2: GALILEO

"I've got a feeling I'm falling"

"No one will be able to read the great book of the
  Universe if he does not understand its language, which
  is that of mathematics."   —*Galileo*

In the preparation for this lab, you read that Galileo first postulated that the **dependent** variable s was proportional to the **independent** variable t. Compare your assessment of Galileo's hypothesis with those of your lab partners to make sure you agree. Let's use the graph you prepared for this lab to check your answer. In Lab 1 you learned that if one variable is proportional to another, the graph of their relationship is a straight line with slope k, the constant of proportionality. In any of your team's graphs of s, do the points appear to lie on a straight line?

Now let's try to reproduce Galileo's work and guess the relationship between s and t. We want to write an algebraic expression in t that turns 1 into 16, 2 into 64, 3 into 144, and so on. That is, when t is 1, the algebraic expression has the value 16; when t is 2, it has the value 64. In function notation, we want to write

$$s = f(t)$$

where $f(0) = 0$, $f(1) = 16$, $f(2) = 64$, and so on. (**Hint:** all values of s in the table—including zero!—are multiples of the same number.)

The **domain** of a function is the set of allowable values for the independent variable. The **range** of the function is the corresponding set of values for the dependent variable. Although our table includes only a few integer values of t, use the formula for s as a function of t that you deduced above and compute $f\left(\frac{5}{2}\right)$.

We think of time as continuous; that is, to get from $t = 1$ to $t = 2$, we pass through all intermediate values of t. In particular, since $1 < \sqrt{2} < 2$, there must be a value of s corresponding to $t = \sqrt{2}$. What is it? Do negative values of t make sense in this physical problem?

In fact, Galileo's **model** for describing the fall of an object from a stationary position works well *only* near the surface of the earth (from the roof of a tall building, for example). Write a description of a reasonable set of values for the domain and range of Galileo's function, explaining your reasons for choosing those values.

Use a smooth curve to connect the points you plotted on your graph. Now use your graphing utility to sketch the function you wrote to model the data in the table. You may need to rename the independent variable x and the dependent variable y to agree with the conventions of your particular calculator or computer program.

The graph on your calculator or computer screen may look very different from the one you drew by hand. There are two reasons for this. We saw in the first lab that changing the scales along the two axes can affect how the graph looks. One way to make the graph on the screen match the one you drew is to experiment with the scales on the two axes. A more important difference is that the graph you drew represented Galileo's **model**: the values for t and s were chosen to make sense in the physical situation that the function modeled. The computer treats your function *abstractly*. It places only those restrictions on the domain and range that are a result of the algebraic operations performed on the independent variable. For example, a computer or calculator (or you!) cannot evaluate

$f(x) = 1/x$ at $x = 0$. (Why not?) For the same reason, you or the computer would have to exclude 2 from the domain of the function $g(x) = 1/(x - 2)$.

Now use your graphing utility to help you determine the domain and range of the *abstract* function relating s and t that you—and Galileo—found. One way to do this is to change the intervals along the horizontal and vertical axes that determine the portion of the graph that appears on the display. We call this operation *changing the viewing window*.

On graph paper, draw the portion of the graph of the abstract function on the interval from $t = -6$ to $t = 6$. With a second color, trace the part of the graph that models the falling-body problem. Use your graph to answer the following:

- When t changes from 1 to 2, what is the resultant change in s?

- When t changes from 2 to 3, what is the resultant change in s?

- When t changes from 3 to 4, what is the resultant change in s?

In Lab 1, you saw that a fixed change in the independent variable C produced a constant change in the dependent variable F. Does a fixed change in the values of the independent variable t result in a constant change in the values of the dependent variable s? How do you think this difference between the functions in Lab 1 and Lab 2 is reflected in their graphs?

For positive t, as t increases, does s increase or decrease? When does s increase more rapidly: when t increases from 1 to 2 or when t increases from 4 to 5? The change in s per unit change in t is called the **average rate of change** of s. Do you think that the average rate of change of s from $t = 9$ to $t = 10$ seconds will be greater or less than the average rate of change of s from $t = 2$ to $t = 3$?

Use your graphing utility to estimate the time at which the object will have fallen 100 feet. At what time will the object have fallen 500 feet?

JACK'S FIELD

The curve on your graphing utility is a **parabola**. It is the familiar and beautiful curve traced out by the trajectory of an arrow. Functions whose **graphs** are parabolas arise in many different situations. Here's one: suppose Jack has 128 feet of fencing to make a rectangular enclosure for his cow. He could fence in a rectangle 10 feet by 54 feet $(10 + 10 + 54 + 54 = 128)$. What would be its area? He could also fence in an area 24 feet by 40 feet. (Check that such a field would require 128 feet of fencing.) What is its area?

The distance around the border of the fenced region is called its **perimeter**. Each of the two fields we've just considered has a perimeter of 128 feet. Draw a sketch of a rectangular field and label one side x. How could you represent the length of the field in terms of x so that its perimeter is 128 feet? Label the other three sides of your diagram and make sure that the perimeter is 128 feet.

Write an expression for the area, A, of the fenced region, measured in square feet, in terms of the length and width as expressions in x. Once a particular width x is chosen, A is completely determined, so x is the independent variable, and $A = f(x)$ is the dependent variable.

Now, here comes the parabola part. Use your graphing utility to sketch $A = f(x)$, then use the graph to help you answer the following questions. (You may have to change the viewing window.)

- What values for x make sense in this physical problem? Explain how you determined your answer.

- What are the corresponding values for the dependent variable A; that is, what is the range of this function, as it applies to the physical problem?

- If Jack wants to enclose the largest possible rectangular area, how should he fence in his field; that is, what dimensions should he use? Describe how you used your graphing utility to find the answer.

THE LABORATORY REPORT

Your lab report should include two pairs of graphs:

- The abstract function $s(t)$ and Galileo's function for falling bodies

- The abstract function $A(x)$ and the area function for Jack's field

Give the domain and range of each of the four functions and explain your answers.

Note the similarities and differences between the graphs of the abstract functions $A(x)$ and $s(t)$. Compare the formula for $A(x)$ to the formula for $s(t)$. What feature(s) account for the similarities in the graphs? What feature(s) account for the differences?

Contrast the rate of change of the dependent variable F in the first lab with the rate of change of the dependent variable s that you examined in this lab. What can you tell about the rates of change for s and F by looking at their graphs? What can you tell about them by looking at the formulas for s and F?

Finally, give the dimensions of the field of largest area that Jack can enclose. Explain how you used the graph of A to determine your answer.

Homework 2.1: "Don't Fence Me In"

1. Recall Jack's field from Laboratory 2. Now suppose Jack has only 90 feet of fencing instead of 128 feet. Following the procedure you used in the lab, write a function for the area A of a rectangular field of perimeter 90 feet in terms of its width x. Use the graph of the function $A(x)$ to determine the dimensions of the rectangular field of greatest area that he can enclose. Does the **shape** of the largest rectangular field depend upon the amount of fencing? Explain.

2. In the space below, show a labeled sketch of the field and a graph of the area function, indicating the point on the graph that gave you both the "best" value for x (that is, the one that yielded the greatest area) and the maximum value for the area.

3. Jill, a mathematics student, decides she will use the same amount of fencing as Jack (128 feet) did to fence in a circular grazing field for the cow. Recall that the **circumference** (perimeter) of a circle is related to its radius r by the formula $c = 2\pi r$, and the area A is a function of r: $A = \pi r^2$. Use these two equations and your calculator to find the area of Jill's field. What was the largest area Jack was able to obtain with 128 feet of fencing? Whose field is larger?

Homework 2.2: Exploring Quadratics

What can the graphing utility help you to discover about quadratic functions? This exercise is designed to encourage you to explore the various possibilities for quadratics. Although you will probably not be able to find exact or complete answers in every case, you will nevertheless be able to use the rapid graphing capability of your computer or calculator to learn much more about quadratic functions than you could if you depended solely upon point plotting.

1. The general expression of a quadratic function takes the form

$$f(x) = ax^2 + bx + c$$

where a can be any value other than zero, and b and c can be any real numbers. You've already worked with quadratic functions such as $f(x) = 16x^2$ ($a = 16, b = 0, c = 0$) and $f(x) = -x^2 + 64x$ ($a = -1, b = 64, c = 0$). Choose several other sets of values for a, b, c and graph the quadratics with your graphing utility. Sketch the basic shape(s) that characterize the graphs of quadratic functions. (Note that you are *not* asked to provide actual graphs, but rather to produce rough sketches of the shapes you saw on the screen.)

2. What do the values of a, b, and c control in the graph of the quadratic function $f(x) = ax^2 + bx + c$? To answer this question, you will need to be more methodical in your investigation, changing only one constant at a time.

   Start with any set of (nonzero) values for a, b, and c. Overlay several graphs: your initial choice for the quadratic function, and your initial choice modified by different values for a.

   (a) How does changing a change the graph?

   (b) What happens to the graph if you change the *sign* of a?

   (c) What happens if you keep the sign of a fixed, but increase the *size* of a?

   (d) Is there any point that doesn't change as you increase the size of a?

3. In a similar fashion, investigate what the coefficient b controls in the graph. Then figure out what effect c has. Write a few sentences about what you have learned.

Mathematicians are very interested in knowing when a function is equal to zero (intercepts the x-axis), is positive (lies above the x-axis), or is negative (lies below the x-axis). Many "nicely behaved" functions can't switch sign without crossing the x-axis first.

4. Write a formula for a quadratic function that crosses the x-axis both at 3 and at -2. (You might want to begin by writing the quadratic in factored form.) Examine how the graph switches from above to below the x-axis (from being positive to being negative). On which interval(s) of the x-axis is your function positive? On which interval(s) is it negative? Illustrate your answer with a quick sketch of the graph.

5. Write a quadratic function whose graph *doesn't* cross the x-axis. Can you factor this function? Does the graph of this function lie above or below the x-axis?

6. Write a quadratic function whose graph *touches* the x-axis but does not cross it. What is special about the factored form of this quadratic?

7. Is it possible for a quadratic function to cross the x-axis in more than two places? Explain why or why not.

Laboratory 3

PREPARATION

Before you begin this lab, let's be sure that you understand function notation. Given a specific formula for a function, say,

$$f(x) = 3x^2 - 2x + 5$$

write the formulas for the following:

$$f(x) + 1, \quad f(x + 1), \quad -f(x), \quad f(-x), \quad f(2x), \quad 2f(x)$$

Repeat this exercise, changing the formula for $f(x)$ to $3\sqrt{x - 1}$. Check your answers with those on page 33.

Function notation is very efficient because it allows us to express the algebraic rule for a mathematical function economically, using only a few symbols. This very economy of expression, unfortunately, can also be a source of confusion. Do you know, for example, the difference in meaning between $f(x) + c$ and $f(x + c)$? between $-f(x)$ and $f(-x)$?

In Lab 2, you studied the quadratic function $16x^2$. You probably plotted several points in order to draw its graph. Suppose that function were modified slightly so that you had to draw $16(x + 3)^2$ or $100 - 16x^2$. Would you want to plot a whole new set of points? Think of all the work! In this lab, you will discover quick methods for graphing functions such as these without point plotting.

In Lab 3, you will imagine yourself a research mathematician. The purpose of the lab is to understand what happens to the graph of any function $f(x)$ when you apply a **transformation** to the function, that is, when you change its algebraic formula in one of the ways you did above. You will have three research tasks during the lab:

- collecting information from your graphing utility

- deciding the effect of the given transformation (backed up by examples)

- determining why the effect occurred

As background for this research, you will need to become familiar with the general shapes of the following functions. (Simply becoming familiar with the general shape of each one will be sufficient.) Sketch their graphs below.

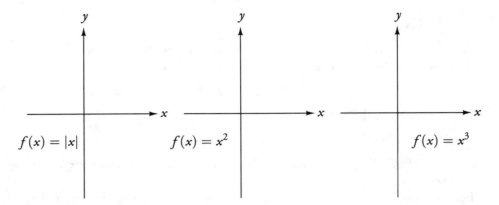

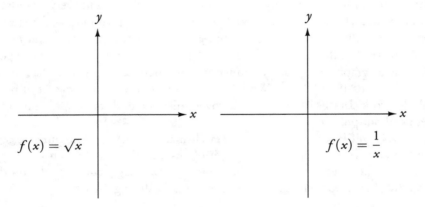

The instructions for the lab give you some specific tasks to perform, but you should not feel restricted to those exercises. As you proceed, you may wonder, What would happen if. . .? Please investigate. The beauty of graphing technology is that it allows us to try out our ideas quickly. If you see something interesting, discuss it with your lab partners. Include in your lab report anything you wondered about, how you explored it, and what you discovered.

LABORATORY 3: EXPLORATIONS—GRAPH TREK

To boldly go where no lab has gone before

Pay particular attention to the title of this lab. The lab instructions are designed to help you discover and understand some mathematical patterns. They aren't meant to inhibit your own creative investigations, however. In fact, the individual explorations of each lab group are an important component of this lab and ought to be included in the report.

SHIFTS

The first transformation to examine is that of adding a constant c to the function $f(x)$, that is, graphing $f(x) + c$. Start with the simplest parabola, $f(x) = x^2$, and draw its graph on your graphing utility. Now overlay the graphs of $y = x^2 + 3$ and $y = x^2 - \sqrt{2}$. Describe what you see; that is, when $f(x)$ is defined as x^2, what does the graph of $y = f(x) + 3$ look like? What does the graph of $y = f(x) - \sqrt{2}$ look like? You should be able to make quick and fairly accurate sketches of these graphs without resorting to point plotting.

Change the formula for $f(x)$. You might use a linear function, another quadratic, or something totally different. Two particularly interesting functions to examine are $\cos(x)$ and $\sin(x)$. (Don't worry yet about the significance of those functions; you'll study them in detail later.) Now watch what happens to its graph when you add a constant. Be sure to use several different constants: not all whole numbers and not all positive. The graphing utility is nothing but a very fast point plotter, and it calculates the points for each new graph without reference to the previous one. *You*, however, can recognize a pattern that enables you to get a new graph from the old one with a minimum of effort. Do several different examples until you are certain you understand the pattern. Describe, in general terms, what happens to the graph of a function when you add a constant to the function.

Suppose the graph of some function $y = CUP(t)$ looked like this:

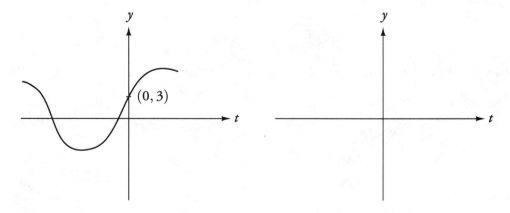

Draw the graph of $y = CUP(t) - 4$. Label its y-intercept.

So far, you've been adding a constant to the function, the dependent variable. Suppose, instead, you were to add a constant to the independent variable; that is, you first add the constant to the input variable, and then perform the function. What would happen to the original graph? From your work in the preparation section, you know that the order of operations makes a difference in the *formula* for the function. [Compare the $f(x) + 1$ and $f(x + 1)$ formulas.] Now you will see that the order of operations also makes a difference in the *graph* of the function.

What do you expect the graph of $y = (x + c)^2$ to look like? (What will be its relationship to the graph of $y = x^2$?) Make your prediction.

Now, graph $f(x) = x^2$ and overlay the graph of $g(x) = f(x + c) = (x + c)^2$. (You choose a value for c.) Is this what you expected? Explain *what* happened and *why*. How does your choice of a value for c affect the basic graph of x^2?

Do you think the behavior you've witnessed is something peculiar to parabolas, or will other functions respond in the same way? Check out your intuition by changing the function. Here are some suggested functions: x^3, $\sqrt{x}$, $\sin(x)$, and $|x|$. Try several examples of adding a constant to the independent variable *before* performing the operation and describe the effect on the graph. (You'll be looking at graphs of such functions as $\sqrt{x - \frac{1}{2}}$ and $|x + \pi|$ and comparing them to the graphs of the simpler functions they most resemble, $\sqrt{x}$ and $|x|$ in this case.)

Suppose you start with $f(x) = x^2 + x$. Adding a constant to the independent variable means looking at, say, $f(x - 1)$. What would the formula be for $f(x - 1)$? How would its graph be related to that of $f(x)$?

Now generalize about the relationship between the graph of some function $BALE(a)$ and that of $BALE(a + c)$. Suppose the graph of $y = BALE(a)$ looks like this:

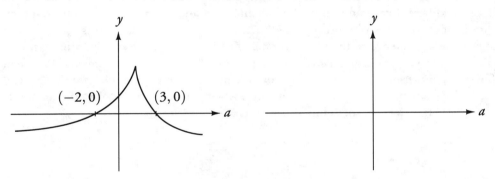

Draw the graph of $y = BALE(a + 2)$. Label its a-intercepts.

By now you should be able to look at the formula for a function and tell whether its graph involves a shift of one of the basic shapes you've been learning. You should be able to predict how far and in what direction(s) the basic shape is shifted. Try these: identify the basic function, predict the kind(s) of shift, then have the graphing utility draw the graph.

(a) $\sqrt{x+2}$ (b) $(x-3)^2 - 5$

(c) $\cos(x+1)$ (d) $x^3 + 2$

(e) $(x+2)^3$ (f) $1 + \cos(x)$

$|x|$ and its relatives have V-shaped graphs. Be sure your interval allows you to see both sides of the V.

(g) $2 + |x - 1|$

The last two graphs each have 2 branches. Choose an interval that shows the whole picture.

(h) $\dfrac{1}{x} - 3$ (i) $\dfrac{1}{(x-3)}$

REFLECTIONS

Another way to get new graphs from old is through reflections. Sketch the graph of $y = -x^2$. [Note that $-x^2$ is different from $(-x)^2$. Check it quickly if you aren't sure.] With the graphs of x^2 and $-x^2$ on the screen at the same time, describe the relationship between the two. Since mathematicians look for patterns, you might expect to find the same relationship between the graphs of any pair of functions $f(x)$ and $-f(x)$—that is, a function and its opposite.

Try graphing several other functions and their opposites. Remember that the opposite of a function is simply the function itself preceded by a negative sign. Here are some suggestions, but you should make up a few of your own as well. Each person in the group should try two or three. Compare your results.

$$\sqrt{x} \hspace{4cm} \sin(x)$$

$$x^2 + 3x - 4 \hspace{1cm} \text{(Remember to take the opposite of}$$
$$\text{the entire quadratic!)}$$

Now it's time to generalize. Suppose the graph of some function $y = SICK(u)$ looks like this:

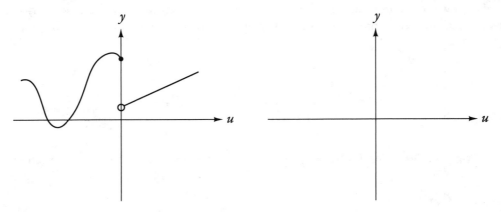

Sketch the graph of $y = -SICK(u)$.

You have been comparing the graph of a function to that of its opposite. Suppose you look instead at $f(-x)$, that is, at the function itself when the input variable is replaced by its opposite. Try this with the function $f(x) = \sqrt{x}$. What's the formula for $f(-x)$? What's the relationship between the graphs of $f(x)$ and $f(-x)$? If you don't see both of them on the screen, pick a different interval. Compare these two graphs with the graph of $-\sqrt{x}$. Try to get all three graphs on the screen simultaneously. The position of the negative sign makes a difference! Explain.

Try graphing $f(-x)$ for several other functions, the same ones you used to study $-f(x)$. Keep doing examples until you can predict with confidence what the graph of $f(-x)$ will look like. [What happens if you start with $f(x) = x^2$? Why?]

Now sketch the graph of $y = SICK(-u)$.

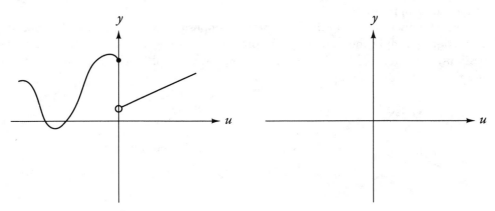

At this point, you should be able to draw the graphs of many different functions by sketching the simple function they most resemble and applying appropriate reflections and shifts. Try these:

$$-\sqrt{x-2} \qquad\qquad 2-\sqrt{x}$$

$$2+\sqrt{-x} \qquad\qquad 5-(x+3)^2$$

$$|-x|-1 \qquad\qquad -\frac{1}{x+1}$$

THE LABORATORY REPORT

Summarize what you learned about the graphs of $f(x)+c$, $f(x+c)$, $-f(x)$, and $f(-x)$ by describing how their graphs are related to the graph of $f(x)$. Tell the effect of each type of transformation and explain why each effect occurred. Illustrate with representative sketches. You do not need to submit every single graph you drew during the lab, but be sure that the graphs that form part of your report make the patterns clear. Do include any interesting or puzzling observations your group may have made.

Answers for preparation exercises

If	$f(x)$	is $3x^2-2x+5$,	If	$f(x)$	is $3\sqrt{x-1}$,
then			then		
	$f(x)+1$	is $3x^2-2x+6$		$f(x)+1$	is $3\sqrt{x-1}+1$
	$f(x+1)$	is $3(x+1)^2-2(x+1)+5$		$f(x+1)$	is $3\sqrt{x}$
	$-f(x)$	is $-3x^2+2x-5$		$-f(x)$	is $-3\sqrt{x-1}$
	$f(-x)$	is $3x^2+2x+5$		$f(-x)$	is $3\sqrt{-x-1}$
	$f(2x)$	is $12x^2-4x+5$		$f(2x)$	is $3\sqrt{2x-1}$
	$2f(x)$	is $6x^2-4x+10$		$2f(x)$	is $6\sqrt{x-1}$

Homework 3.1: Vertical Stretching and Compression

$$c \cdot f(x)$$

In this assignment, you'll look at another way to modify a function and watch what happens to its graph. In order to get accurate pictures, you should have equal scales on the horizontal and vertical axes. Adjust your graphing utility so that the x- and y-axes have the same scale, and keep it that way for this entire assignment.

1. Go back to good old x^2. Overlay the graph of $3x^2$. Sketch. Describe what the 3 does to the graph.

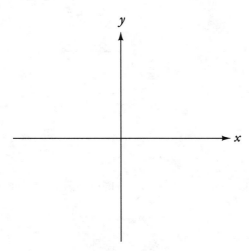

2. Now overlay the graph of $\frac{1}{2}x^2$. Sketch. Describe the effect of the $\frac{1}{2}$.

Examine $c \cdot f(x)$ for several other functions. (But don't feel that these are the *only* ones you may explore.)

3. Compare $3\sqrt{x}$ and $0.3\sqrt{x}$ to $\sqrt{x}$. Make sketches. Identify each graph.

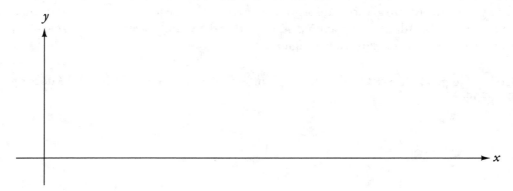

4. Compare $2\cos(x)$ and $\frac{1}{2}\cos(x)$ to $\cos(x)$. Make sketches. Describe what you see.

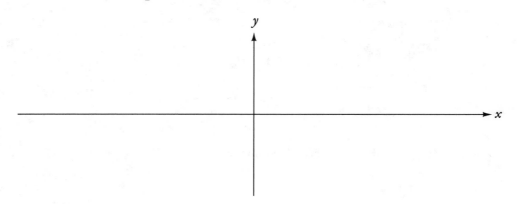

5. Do one more set of your own choosing.

    (a) Pick a function $f(x)$

    (b) Sketch its graph

    (c) Sketch two more graphs, $a \cdot f(x)$ and $b \cdot f(x)$, where a and b are constants, $a > 1$ and $0 < b < 1$.

    (d) Write a couple of sentences to generalize the effect on the graph of a function of multiplying the function by a positive constant.

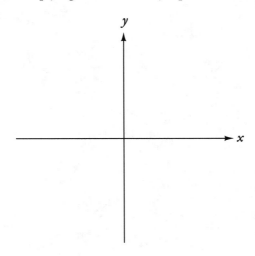

Now examine $c \cdot f(x)$ where c is negative. You already have an idea, from your work in Lab 3, of what might happen. Try a couple, such as $-3 \cdot f(x)$ and $-\frac{1}{2}f(x)$. In a sentence or two, generalize what you see.

Homework 3.2: Horizontal Stretching and Compression

$$f(c \cdot x)$$

Here's yet another way to modify a function algebraically. In this assignment, you'll see that multiplying the independent variable by a constant will have a predictable effect on the graph of the function. The best functions to use here are $\cos(x)$ and $\sin(x)$, even though you may not yet understand their significance, because they clearly show the difference between multiplying a function by a constant (first perform the function, then multiply) and multiplying the independent variable by the constant (first multiply the input variable, then perform the function).

Adjust your graphing utility so that the scales on the horizontal and vertical axes are equal. This ensures that the shapes of the graphs will not be distorted. Recall, from the previous assignment, what happened to the graph of $\cos(x)$ when you changed it to $2\cos(x)$.

1. Now try $\cos(2x)$. Show by means of a sketch that the 2 in $2\cos(x)$ *stretches* the graph of $\cos(x)$ vertically by a factor of 2, whereas the 2 in $\cos(2x)$ *compresses* the graph of $\cos(x)$ horizontally by a factor of 2. (Show several complete cosine waves.)

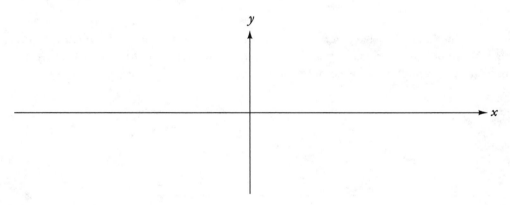

2. What would you expect of $\cos(\frac{1}{2}x)$? Graph $\frac{1}{2}\cos(x)$, $\cos(x)$, and $\cos(\frac{1}{2}x)$. Describe what you see.

3. (a) Draw $\cos(x)$ using the interval $-4 \leq x \leq 4$.

(b) Draw $\cos(2x)$ using the same interval.

(c) Now, redraw $\cos(2x)$ using the interval $-2 \leq x \leq 2$.

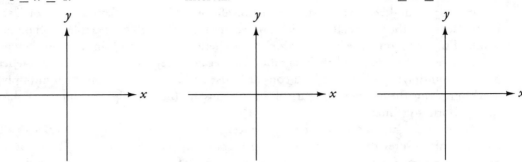

Two of these graphs look the same. How are they different?

4. Will the sine function behave in the same manner? Choose a value for c (something other than 2); graph $\sin(x)$, $c \cdot \sin(x)$, and $\sin(c \cdot x)$. Give a representative sketch, using an interval wide enough to show complete waves.

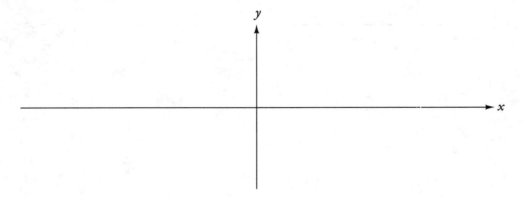

5. Let $f(x)$ be the function $x^3 - x$. Sketch its graph in the vicinity of its x-intercepts. On the same axes, sketch the graphs of $2f(x)$ and $f(2x)$. Provide a scale. (You may need to play around with the interval to get a good picture. Get close to the intercepts. Each graph has two loops; be sure you see them.) What effects do you see: stretching or compression? vertical or horizontal?

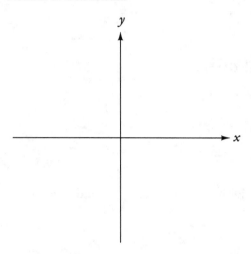

THE SILLY PUTTY APPROACH TO GRAPHING

6. The effects you've been seeing apply to all functions, although it's often difficult to tell whether a graph has been stretched vertically or compressed horizontally. Take, for example, the function $f(x) = x^2$ and consider the vertical stretch $9 \cdot f(x)$. What value of c, in a horizontal compression $f(c \cdot x)$, will result in an identical graph?

7. Sometimes, a horizontal stretch can look like a vertical stretch. Consider the function $g(x) = \frac{1}{x}$ and a horizontal stretch, $g(\frac{1}{4}x)$. What vertical stretch $c \cdot g(x)$ will produce the same graph?

Multiplying a function by a constant has one effect on its graph; multiplying the independent variable by a constant has a different effect (even though, as you saw above, the results occasionally can be similar). After finishing this assignment and the preceding one, you should understand both effects.

To check your progress, answer the following questions about these functions:

$$1 - \sin(0.2x), \quad 60\sin\left(x - \frac{\pi}{4}\right), \quad \frac{2}{5}\sin(x + 1), \quad \sin(3x) + 4$$

8. Which of the functions has a constant that behaves as a trash compactor, squashing the sine graph toward the x-axis?

9. Which has a constant that grabs the tops and bottoms of the sine wave and tugs them vertically, away from the x-axis, as in stretching a rubber sheet?

10. Which has a constant that stretches $sin(x)$ horizontally away from the y-axis, as with an overextended accordion?

For further exploration, you might look at what happens to $f(c \cdot x)$ when the constant c is negative.

Homework 3.3: Symmetry about the Y-axis

("even" symmetry)

Beauty is truth, truth beauty,—that is all
Ye know on earth, and all ye need to know.
"Ode to a Grecian Urn," John Keats

Beauty is often associated with symmetry. We would all agree that the shape of a Grecian urn is very symmetrical. When the curve representing the graph of a function $f(x)$ is placed symmetrically on the coordinate axes, we say it is **symmetric about the y-axis** because, if you folded this paper along the y-axis, the two halves of the graph would lie on top of each other. (Try it!)

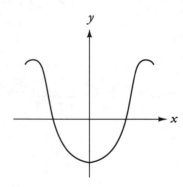

1. Here are the graphs of three more functions, $f(x)$, that are symmetric about the y-axis. For each of these functions, draw the graph of $f(-x)$. [Recall that, in Lab 3, you learned how to draw the graph of $f(-x)$, given the graph of $f(x)$. $f(-x)$ is *not* an upside-down version of $f(x)$.]

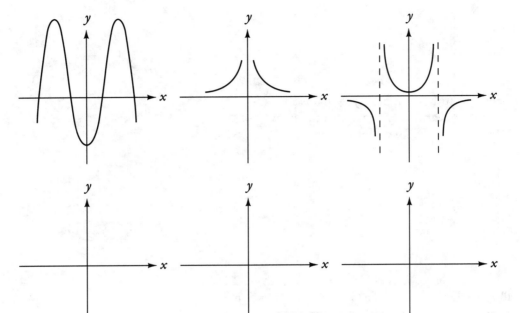

What do you observe about the three graphs you just drew?

Here are the formulas for the three functions.

(a) $f(x) = -x^4 + 10x^2 - 9$

(b) $f(x) = \dfrac{1}{x^2}$

(c) $f(x) = \dfrac{1}{4 - x^2}$ (**Warning:** remember, this really means $\dfrac{1}{4 - (x^2)}$.)

2. For each of the functions above, write the formula for $f(-x)$. Compare it to the formula for $f(x)$. What do you observe? Recall from Lab 3 how the graphs of $f(x)$ and $f(-x)$ are related. In your own words, explain why you should expect the algebraic results you just obtained, simply by looking at the pictures.

In fact, the formula for a function reveals whether or not its graph will have y-axis symmetry. Whenever the formula for $f(-x)$ is the same as the formula for $f(x)$, the graph of the function will be symmetric about the y-axis.

3. Test the following, algebraically, to predict whether or not their graphs will show y-axis symmetry. Check your results by looking at their graphs.

(a) $7 - 3x^2$ (b) $x^6 - 40x^2 + 5$

(c) $x^6 - 40x^3 + 5$ (d) $\dfrac{x}{x^2 - 4}$

(e) $\dfrac{x^2}{x^2 - 4}$

Homework 3.4: Symmetry about the Origin

("odd" symmetry)

The curve here mimics the symmetry of the galaxy above. It is placed on the coordinate axes symmetrically, but its symmetry is different from the even symmetry of the Grecian urn curve in the previous exercise. One way to describe its symmetry is to imagine placing a pin through the graph at the origin and then spinning or rotating the paper 180° around the pin, as a pinwheel (or a galaxy) rotates about its center. The new position of the graph is identical to its old position. We say such a graph is **symmetric about the origin**.

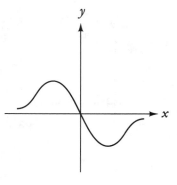

1. Here are the graphs of three more functions, $f(x)$, that are symmetric about the origin. Below each one, draw the graph of $f(-x)$. Below that, in the third row, draw the graph of $-f(x)$. (Be sure to use the *original* $f(x)$ each time.)

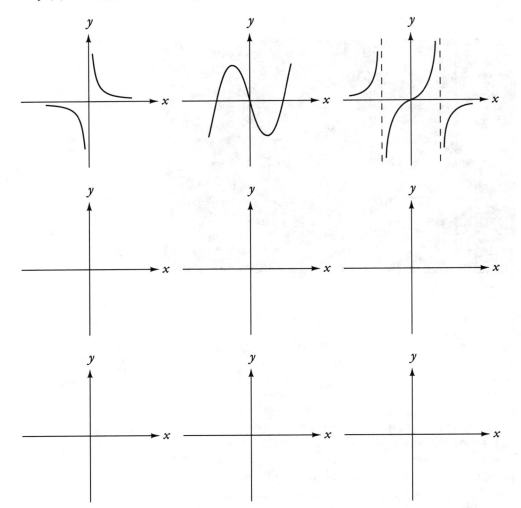

2. Examine the six graphs you just drew. What do you notice? In each case that you drew the graph of $f(-x)$, you needed to reflect the graph of $f(x)$ about one of the axes. In each case that you drew the graph of $-f(x)$, you also needed to reflect the graph of $f(x)$ about one of the axes.

   (a) The graph of $f(-x)$ is the reflection of the graph of $f(x)$ in the _____ axis.

   (b) The graph of $-f(x)$ is the reflection of the graph of $f(x)$ in the _____ axis.

Here are the formulas for the three functions.

(a) $f(x) = \dfrac{1}{x^3}$

(b) $f(x) = x^3 - 4x$

(c) $f(x) = \dfrac{x}{4 - x^2}$

3. For each of the functions above, write the formula for $f(-x)$ and the formula for $-f(x)$. Using rules of algebra, show that in each case these two formulas are the same. In your own words, explain why you could have expected such results simply by looking at the three given graphs.

In fact, the formula for a function is a giveaway for this type of symmetry, too. Whenever $f(-x)$ is *algebraically equivalent* to $-f(x)$ (that is, when we can make them the same using rules of algebra), the graph of $f(x)$ will be symmetric about the origin.

4. Test the following, algebraically, to predict whether or not their graphs will show origin symmetry. Check your results by looking at their graphs.

(a) $7x - 3x^3$

(b) $x^5 - 25x^3 + 100x$

(c) $x^5 - 25x^3 + 100$

(d) $\dfrac{3}{x - 3}$

(e) $\dfrac{x}{x^2 - 4}$

(f) $\dfrac{x^2 - 4}{x}$

5. Does a graph with origin symmetry necessarily pass through the origin? Explain.

6. Suppose the graph has origin symmetry *and* a *y*-intercept. Does it pass through the origin? Explain.

Homework 3.5: Symmetry Test

Let's summarize what we learned in the previous two homework lessons.

If we know the formula for a function, we can test for y-axis or origin symmetry by evaluating $f(-x)$.

- If $f(-x)$ is algebraically equivalent to $f(x)$, the graph has symmetry about the y-axis (also known as "even" symmetry).

- If $f(-x)$ is the opposite of $f(x)$ [that is, if $f(-x)$ and $-f(x)$ are algebraically equivalent], the graph has symmetry about the origin ("odd" symmetry).

- If $f(-x)$ is something else, the graph has neither type of symmetry.

1. Apply the symmetry test to the following functions. Predict whether their graphs will have y-axis symmetry, origin symmetry, or neither. Then check your results by looking at the graphs.

   (a) $r^4 + \pi$ (b) $r^2 + \pi r$

   (c) $\sqrt{t}$ (d) $\sqrt[3]{t}$

   (e) $2 - s + s^3$ (f) $\dfrac{x}{x - 3}$

   (g) $\sqrt{x^4 - 9}$ (h) $\dfrac{s}{3 - s^2}$

   (i) $\dfrac{z^2 - 9}{z^2 + 2}$ (j) $\dfrac{\sqrt[3]{t}}{t^3}$

Homework 3.6: Absolute Value in Functions

By now you should be getting the idea that success in graphing depends in large part upon learning a few basic shapes and then being able to recognize patterns. This assignment will teach you how to toss absolute values into the mix.

$|f(x)|$ THE ABSOLUTE VALUE OF A FUNCTION

The absolute value of a quantity is always nonnegative. The absolute value of a function, therefore, will have to be nonnegative as well.

1. Have your graphing utility draw both $f(x) = x^2 - 4$ and $g(x) = |x^2 - 4|$ on the interval $-3 \leq x \leq 3$. Sketch what you see, using two different colors and showing clearly where the graphs coincide and where they differ.

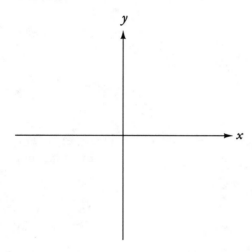

Explain what the absolute value did to the graph of the second function. Why did some portions of the graph change, while others remained the same?

2. In the same manner, graph $f(x) = 4 - x^2$ and $g(x) = |4 - x^2|$. Compare your results with what you saw in the first set of graphs.

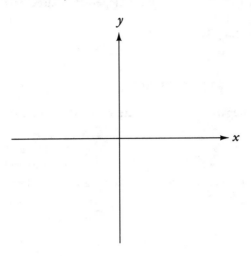

3. For each function given, sketch a quick graph. Then, write the formula for $g(x) = |f(x)|$ and use your sketch to make a graph of $g(x)$. Use your graphing utility to check your work.

(a) $f(x) = x - 2$ (b) $f(x) = \frac{1}{x}$ (c) $f(x) = \sin(x)$

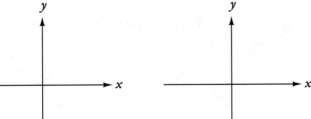
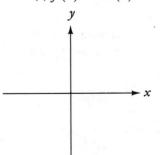

4. If the graph of some function $KIWI(w)$ looks like the graph below, sketch the graph of $|KIWI(w)|$.

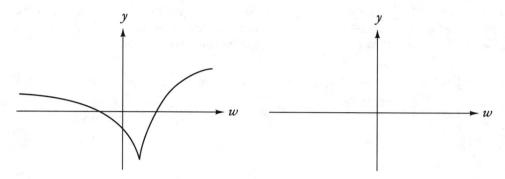

5. Write a sentence or two summarizing what you have learned about taking the absolute value of an entire function. In the next section you will learn what happens to a function when the absolute value is applied first.

$f(|x|)$ THE FUNCTION OF AN ABSOLUTE VALUE

Something different happens if we apply the absolute value to the **independent** variable. The expression $f(|x|)$ implies that, regardless of whether the input variable is negative or positive, the function will use only the **magnitude** (a nonnegative quantity) of that variable. For example, if x happens to be -5, the value to be computed will be $f(|-5|)$, or $f(5)$.

6. Let's see how this affects the graph. Draw the graph of $f(x) = x^2 - 2x$ on the interval $-4 \leq x \leq 4$. Write the formula for $g(x) = f(|x|)$.

7. Overlay the graph of $g(x)$. Sketch what you see, using two colors and showing clearly where the graphs coincide and where they differ.

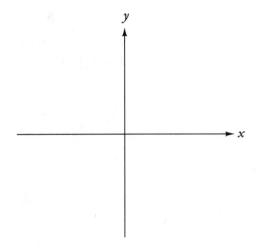

Next to the graph you drew, **describe** what happened, and **explain** why the graph of $f(|x|)$ looks the way it does.

8. You should observe that some of the outputs are still negative (that is, below the x-axis). Explain why, even though we're dealing with an absolute value, we can still end up with negative values for the **function**. Why didn't that occur in the last section, when you took the absolute value of a function?

9. Here are some practice problems. For each function given, sketch a quick graph. Then, write the formula for $g(s) = f(|x|)$, and use your drawing to make a graph of $g(x)$. Use the graphing utility to check.

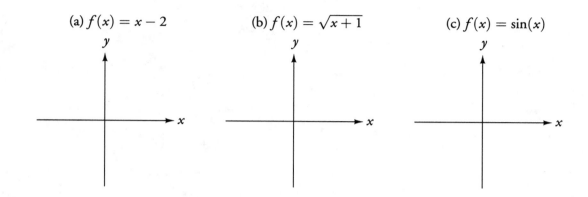

(a) $f(x) = x - 2$ (b) $f(x) = \sqrt{x + 1}$ (c) $f(x) = \sin(x)$

10. Using the graph of $KIWI(w)$ on page 55, sketch the graph of $KIWI(|w|)$.

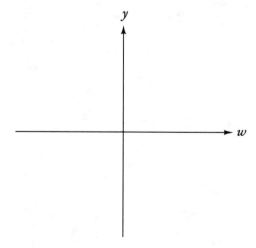

11. Explain what you did, in each case, to the graph of $f(x)$ to produce the graph of $f(|x|)$.

Laboratory 4

PREPARATION

It is a hot day in the middle of August. You are typing away on your computer (things are going great) and all of a sudden . . . the power drops! Hours of work are down the drain (you never save as you type!).

During times of prolonged extreme temperatures, New England electric companies are not always able to keep up with demand. Blackouts and brownouts occur, causing major and minor power disruption. In order to provide better service to their customers, electric companies routinely investigate the relationship between power consumption and temperature.

Utility companies also supply data to customers to help customers recognize patterns of electrical usage in their homes or businesses. Below are two examples of the information supplied to homeowners in Massachusetts.

House 1: Customer printout from Northeast Utilities; home has an air conditioner.

House 2: Customer printout from Northeast Utilities; home has electric heat.

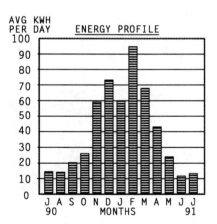

Think about the effect that temperature has on the power usage for these two homes. How is power consumption affected in the first house when summertime temperatures soar? As temperatures drop in the winter, what happens to the electric bill for the second house?

Hot spells in New England bring warnings to customers to cut back on electrical usage or face electrical brownouts. ("Turn your air conditioners off for 15 minutes every hour. Set air conditioner thermostats at 75°F instead of at 70°.") What effect do you expect the higher summer temperatures to have on electrical power consumption? Draw a graph illustrating your impression of the way in which **power usage** is related to **temperature** during the warmer weather of late spring and summer.

Which is the dependent variable? the independent variable? What values did you use as reasonable temperatures for spring/summer weather? How did you decide whether the shape of your graph should be a line or a curve? In plain English, and in the context of electric power usage, what's the difference between a straight-line graph and a curve? Discuss these questions with your partners when you get to the lab.

Extreme cold weather also poses problems of increased demands for power. Speculate on what a graph of the relationship between power consumption and temperature might look like during a New England winter.

Now piece together your graphs to get a single graph that summarizes the relationship between temperature and power consumption. For the temperature axis, you will need to decide upon a set of reasonable temperatures that might occur in New England over the course of a year. (How low do you think the temperature might go? how high? These numbers will determine the interval on your temperature axis.) You're unlikely to have any actual values for power consumption, so you may omit a scale on the power consumption axis. Bring a sketch of this graph to the lab so that you can compare results with your partners.

LABORATORY 4: ELECTRIC POWER

"You light up my life"

Using data supplied by Massachusetts Electric Company, real-life models predicting monthly power consumption based on average monthly temperature have been constructed for your use in this lab.

MODEL 1: PIECEWISE LINEAR MODEL

If you had decided that the model relating electrical power consumption to temperature was a V-shaped graph of two lines pieced together, one representing the relationship during the warm weather and the other for the cooler weather, the model partially specified below may be close to the shape of the graph you drew in preparation for this lab.

For the cooler temperatures use:

$$p_1(x) = 1588 - 8.7x$$

and for the warmer temperatures use:

$$p_1(x) = 389 + 11.4x$$

where x is the average monthly temperature as measured in degrees Fahrenheit and $p_1(x)$ is the resulting power consumption as measured in millions of kilowatt hours.

Note: There are differences between the graphs you drew in the preparation section and the two graphs you will study in this lab. One important difference in the models presented here is that the only temperatures used are average monthly temperatures (the average, over the entire month, of the daytime and nighttime temperatures), so you probably won't look at temperatures as extreme as those you may have considered earlier. Another difference is that, in the preparation section, you had no way to measure the power consumption; you could indicate on your graph only that the power usage for one temperature was higher or lower than the power consumption for a different temperature. Nevertheless, the general shape of the graph you drew should be comparable to one of the models presented here.

Graph the two lines that make up $p_1(x)$. At what temperature do these lines meet? Complete the algebraic definition of the model by filling in the intervals below.

$$p_1(x) = \begin{cases} 1588 - 8.7x & \text{for } x \text{_____} \\ 389 + 11.4x & \text{for } x \text{_____} \end{cases}$$

In terms of what happens in the real world, explain why the coefficient for x is negative in one portion of Model 1 and positive in the other portion. (Think about what the slopes of the lines signify in the context of temperature and power usage.)

MODEL 2: PARABOLIC MODEL

If you decided that the relationship between power consumption and temperature could best be represented by a curve, your sketch may have more closely resembled the graph of the model specified below.

$$p_2(x) = 1953 - 28.7x + .251x^2$$

Comparing the two models

Graph $p_1(x)$ and $p_2(x)$ on the same viewing screen. [If your graphing utility won't piece together the line segments of $p_1(x)$ for you, you will have to keep track of which line is the appropriate one to be using for which temperatures.]

Suppose the average monthly temperature for July were 68° F. What would you estimate power usage to be using Model 1? Model 2? Find this information from your graph. Then, find the answer using algebra.

For what temperatures do the two models predict the same power consumption? Solve this problem using your graph. Now write down the algebraic equivalent of the question and use algebra to check the solution. (You will need your calculator and the quadratic formula.)

For what temperatures does Model 1 predict higher power consumption than does Model 2? Write down the algebraic equivalent of this question. [How did you handle the fact that $p_1(x)$ consists of two pieces?]

Restrict your viewing screen so that temperatures range from 16°F to 77°F. (These are minimum and maximum average monthly temperatures recorded over a ten-year period.) How different from one another do the two models appear over this range of temperatures? Over what temperature interval would you say the two models exhibit the most discrepancy in predicting power consumption?

The Difference Function

When $x = 68°$ F, what was the discrepancy in power usage between Models 1 and 2? You have just computed $p_1(68) - p_2(68)$, the **difference** between the two models when $x = 68$. To compare Model 1 with Model 2 for other temperatures, you can form the function

$$d(x) = p_1(x) - p_2(x)$$

(Aren't you glad you don't have to do the subtraction for each temperature individually?)

Write down the algebraic expression for $d(x)$. [Be careful. Since $p_1(x)$ is a piecewise function, $d(x)$ will need to be defined as a piecewise function also.] Sketch the graph of $d(x)$.

Use the graph of $d(x)$ to find the average monthly temperatures for which the two models predict the same power consumption. Compare the answer you obtained using $d(x)$ with your answer computed using the graphs of the two models. How can you use the graph of $d(x)$ to determine the temperature interval(s) for which Model 1 predicts higher power usage than does Model 2? Verify that this approach yields the same solution that you got before.

THE LABORATORY REPORT

This lab presented two mathematical models for predicting power usage.

- Explain how each model represents the reality. (What's the significance of the shape? What does each one tell you about patterns of power usage, as affected by temperature?)

- Discuss their similarities and differences. (Be sure to explain, in terms of power usage, the difference in meaning between the straight line and the curve.)

- Show how you used the graphs to estimate the solutions to an equation and an inequality and explain the connection between those graphical solutions and the solutions you obtained using algebra.

- Explain how you used the difference function $d(x)$ to compare the two models.

- Illustrate your discussion with clearly labeled graphs.

- Include, if you wish, comments on the graph you drew as preparation.

Homework 4.1: Inequalities

"One of these things is not like the other . . . "

—*Sesame Street song*

SOLVING INEQUALITIES

In Lab 4, "Electric Power," you solved inequalities in the context of comparing models for power consumption. Suppose, as a mathematician, you are interested in comparing two functions: a linear function

$$f(x) = 2x + 4$$

and a quadratic function

$$g(x) = x^2 - 3$$

Here is a question of interest to mathematicians: For what values of x does the linear function produce larger values than the quadratic function?

1. Translate this question into an algebraic inequality.

2. Graph the two functions and answer the question using your graphing utility. Now use the inequality and find an answer to this question using algebra. (You will need the quadratic formula.) Which process produces an exact answer? Which process produces an approximate answer?

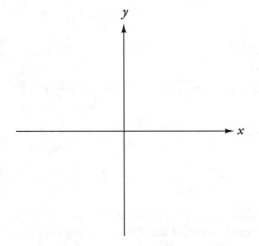

Every precalculus textbook contains exercises similar to the one given below.
   Find the solution set to the inequality

$$x^3 - 3x^2 + 1 < 4x + 1$$

Is it possible to use your graphing utility to solve this problem? First you need to translate the algebraic problem into a statement about two functions. (This is just the reverse of what you did for the first problem.)

$$\text{Let } f(x) = x^3 - 3x^2 + 1 \text{ and } g(x) = 4x + 1$$

   Solving the given inequality is equivalent to answering the following question: for what values of x is $f(x) < g(x)$?

3. Graph the two functions and sketch the graph on paper. Explain how you can use the graph to solve the inequality.

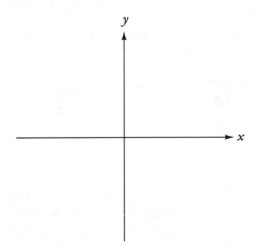

4. Now use your graphing utility to find the solution to your question. Is this solution an exact solution or an approximate solution?

5. Go back and solve the same inequality algebraically. (First, change it into an equation. Solve for x **exactly**. Now, look at the graph of the two functions. These values of x are the *only* values for which the graphs intersect. On which intervals is the graph of $f(x)$ below the graph of $g(x)$?)

6. Compare this solution to the one you got using your graphing utility.

Using your graphing utility to check your answer

7. Solve the following inequality algebraically, and then use your graphing utility to check the solution.

$$-2x^2 + 6 > x^2 - 7$$

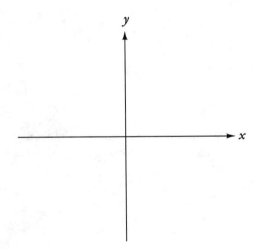

Finding solution sets

8. Find the solution sets to the following inequalities. You choose the method (your graphing utility or algebra), but state whether your solution is exact or approximate.

(a) $3x^3 - 4x^2 > x - 5$

(b) $3\sin(x) < 2x - 1$

(c) $x^2 + 5 < 0$

(d) $\frac{3}{2}(x - 1) \geq \sqrt{x} + 2$

(e) $2x^2 + 3x - 4 < 4x - 1 < x^2 + x$

The solution to (e) is the interval on which both of the inequalities are true simultaneously, in other words, the x-values for which the first parabola is on the bottom, the line in the middle, and the second parabola on the top.

Homework 4.2: "It's Not Easy Being Green"

Polyethylene film is often used for greenhouse coverings in Japan. Manufacturers of the film would like to produce as thin a film as possible so they can produce more square feet of material for less money. The customers (greenhouse owners) want a film thick enough to resist tearing from wind and rain, but not so thick that the light that passes through is insufficient to grow healthy plants. A film more than 2 mm thick will be too opaque for use as a greenhouse covering.

Assume that the cost to the consumer can be modeled by the function

$$C_1(x) = .3x^2 - 1.2x + 6.2$$

and the cost to the manufacturer by the function

$$C_2(x) = .5x^2 + .4x + 4.08$$

where x represents the thickness of the film in millimeters and where the cost is measured in units of $100,000.

1. Film thicknesses for which this problem makes sense lie in the interval from 0 to 2 mm. Graph the two cost functions, restricting the x values to the interval $(0, 2)$. As the thickness x increases, what happens to the cost to the consumer? What happens to the cost to the manufacturer? [Mathematicians would apply the terms **decreasing** and **increasing** over $(0, 2)$ to the functions $C_1(x)$ and $C_2(x)$, respectively.] For what thicknesses is the cost to the producer less than the cost to the consumer?

2. Japanese business philosophy (quite different from that in the United States) concerns itself with making a quality product that suits both the manufacturer and the consumer. The government sets a standard thickness for the film, selecting the thickness that will minimize the **loss**, or total cost, to society. One way to assess the loss to society is to form a new function, $L(x)$, by taking the sum of the cost to the consumers and the cost to the manufacturers. Graph $L(x)$, the function representing the loss to society, and determine the standard thickness that the government will require.

3. If the producer decides to cheat on the standard just a bit and produces a film that is .45 mm thick, how much will the company save? How much will the consumer lose? By how much will society be cheated? [Note that the amount by which society is *cheated* is not simply the value of $L(x)$ for $x = .45$ mm, but rather the difference between that value and the loss to society that should have been expected.]

   Don't round off your results. Use all the accuracy your computer or calculator provides and express your answers in **dollars**, recalling that C_1, C_2, and L are measured in $100,000 units.

Homework 4.3: The Least Squares Line

MODEL SELECTION

In Lab 4, "Electric Power," two models summarizing the relationship between power consumption and average monthly temperature were presented. Where did these models come from? Below you will find a subset of the actual data from Massachusetts Electric Company, rounded to integer values. In the x column are average monthly temperatures measured in degrees Fahrenheit, and in the y column are energy consumption values in millions of kilowatt hours.

x	y
63	1048
64	1110
67	1134
69	1198
72	1155
76	1312

1. First plot these points and then sketch a line that you think best summarizes the data you plotted. (You won't be able to force the line to pass through all of the points.) Find an equation for the line that you have drawn. After you fill in Table A, compare the power consumption you would predict by using your line and the actual power usage. The predicted y values are the results of plugging the six x values (from the table) into your linear equation.

Table A

x	error $y - \text{predicted } y$	$(\text{error})^2$ $(y - \text{predicted } y)^2$
63	1048 − _____ =	
64	1110 − _____ =	
67	1134 − _____ =	
69	1198 − _____ =	
72	1155 − _____ =	
76	1312 − _____ =	

2. In picking the best line for the data, statisticians often select a line in which the errors cancel each other out. How close did you come to selecting a line in which the sum of the errors = 0 (that is, entries in the error column sum to 0)? How could you adjust your line to get this sum closer to zero? (Would you move your line up or down? make the line tilt more or less?)

The sum of the squared errors, SSE [you will have to sum the entries in the (error)^2 column], is also used for selecting a line that "best" describes the data. The line that has the smallest SSE is called the **least squares line** (also referred to as the estimated regression line) and is a common choice for modeling relationships when a plot of the data falls roughly along a line. Find the SSE for your line.

If you have a calculator that can compute the regression line, figure out how to enter the data and have the calculator tell you the slope and intercept of the "best" fitting line to your data. If you do not have access to such a calculator, follow the procedure listed below.

CALCULATING SLOPE AND INTERCEPT OF THE LEAST SQUARES LINE

Compute the average of the x's; call this value $\bar{x}$.

Compute the average of the y's; call this value $\bar{y}$.

To compute the slope and y-intercept of the least squares line, fill in the entries in Table B. You will need to sum the entries in the second and fourth columns. Those sums will be designated as $SUM(x - x)^2$ and $SUM(y - y)^2$, respectively.

Table B

Deviations of x's from their average $x - \bar{x}$	Squared entries of column 1 $(x - \bar{x})^2$	Deviations of y's from their average $(y - \bar{y})$	Squared entries of column 3 $(y - \bar{y})^2$
$63 - \bar{x} = $ _____		$1048 - \bar{y} = $ _____	
$64 - \bar{x} = $ _____		$1110 - \bar{y} = $ _____	
$67 - \bar{x} = $ _____		$1134 - \bar{y} = $ _____	
$69 - \bar{x} = $ _____		$1198 - \bar{y} = $ _____	
$72 - \bar{x} = $ _____		$1155 - \bar{y} = $ _____	
$76 - \bar{x} = $ _____		$1312 - \bar{y} = $ _____	

The **slope** of the least squares line is

$$b = \frac{SUM(y - \bar{y})^2}{SUM(x - \bar{x})^2}$$

The *y*-**intercept** of the least squares line is

$$a = \bar{y} - b\bar{x}$$

(**Note:** Whereas mathematicians generally use m to designate the slope of a line and b as the y-intercept, statisticians often use b for the slope of the regression line and a to designate the y-intercept of this line. Since most calculators programmed to compute the regression line use this latter notation, we have chosen to use this notation here.)

3. Write the equation of the least squares line.

4. Overlay the graph of the least squares line on your plot of the data. Does this line appear to summarize the general flow of the plotted points? How do the slope and intercept of the least squares line compare to the line that you selected? (Is the least squares line tilted more or less steeply than your line?)

5. Set up a table similar to Table A with columns of errors and squared errors for the least squares line. Is the sum of the errors close to 0? Compute the SSE for the least squares line. How does it compare to the SSE of the line that you selected? Why do you think that statisticians would prefer the line with the smaller SSE?

6. Compare the model that you just created when you computed the least squares line to the warm weather portion of the piecewise linear model $p_1(x)$ in Lab 4. [$y = 389 + 11.4x$ is the warm weather portion of $p_1(x)$.]

Two suggestions for making this comparison are provided below.

Suggestion 1: Overlay the graphs of the two models using a viewing screen where x ranges from 63 to 76, the range of temperatures in your data set. How different do the two graphs look? At what temperature does your model predict the same power usage as $p_1(x)$? What is the largest discrepancy in predicted power usage between these two models when you restrict x to be from 63 to 76?

Suggestion 2: Graph the difference, $d(x)$, between the two linear functions using a viewing window with $63 \leq x \leq 76$. At what temperature would the two models predict the same energy usage? What is the largest difference in predicted power consumption between these two models?

Actual data for checking warm weather model

For students who wish to check the calculations used for the warm weather portion of the piecewise linear model (from Lab 4), the actual data is given below. (Data have been rounded to two decimal places and is from the months June, July, August, and September of the years 1986–1989.)

Average monthly temperature	Power consumption
63.23	1048.40
68.48	1055.94
71.81	1088.53
64.83	1081.63
62.77	1121.68
64.39	1110.17
72.03	1155.41
66.87	1134.52
63.53	1166.57
71.29	1215.95
75.61	1312.37
68.70	1198.38
64.13	1193.52
70.42	1206.26
72.58	1267.14
68.30	1254.12

Laboratory 5

PREPARATION

In Lab 1, you examined the function $F = \frac{9}{5}C + 32$, which describes how degrees Fahrenheit depends upon degrees Centigrade. In Lab 2, you saw that the function $d = 16t^2$ showed that the number of feet an object had fallen depends upon the number of seconds it has been falling.

But many quantities are determined by several inputs rather than a single one. The elevation of a point on the surface of the earth, for example, depends on two independent variables, longitude and latitude. And the price at the pump of a gallon of gasoline in this country depends upon the wholesale price, the dealer markup, the federal tax rate, and the state tax rate. (Each of these quantities, in turn, depends upon many other factors, but that's another story.)

In the space below, write two of your own examples of functions that depend on two or more independently varying quantities. You may or may not be able to give formulas for these functions.

The area of a square depends only on the length of a side. We write $A(s) = s^2$, the notation $A(s)$ emphasizing the dependence of A on a single variable s. The area of a rectangle depends on both the length and the width. We write $A(l, w) = l \cdot w$, emphasizing the dependence of A on the **two** variables l and w. You are already using such function notation for functions of a single variable; now we wish to extend it to functions of several variables.

77

Here are some diagrams that appear inside the front cover of a typical precalculus text:

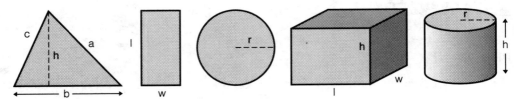

Write the area functions for the triangle, the rectangle, and the circle, expressing them in function notation. How many independent variables appear in each function?

Write the perimeter (circumference) functions for the same shapes. How many variables does each function have?

For the two solids, write the surface area and volume formulas, expressing them in function notation. How many independent variables occur in each?

Note: Bring these pages with you to the lab.

LABORATORY 5: BORDEAUX

"Roll out the barrel . . . "

The following excerpt is from a story in the *New York Times* (March 4, 1990), describing the furor among wine critics over an economist's attempt to quantify the determination of the quality of a particular vintage.

Wine Equation Puts Some Noses out of Joint

Calculate the winter rain and the harvest rain (in millimeters). Add summer heat in the vineyard (in degrees centigrade). Subtract 12.145. And what do you have? A very, very passionate argument over wine.

Professor Orley Ashenfelter, a Princeton economist, has devised a mathematical formula for predicting the quality of red wine vintages in France. And the guardians of tradition are fuming.

. . . It is widely agreed that weather influences wine quality. What few understand, Ashenfelter argues, is that a mere handful of facts about the local weather tell almost all there is to know about a vintage.[1]

The Bordeaux Equation, cited in the article, that Professor Ashenfelter uses to predict wine quality is

$$Q = 0.0117WR - 0.00386HR + 0.6164TMP - 12.145$$

where

$$WR = \text{winter rain (October through March) in millimeters}$$

$$HR = \text{harvest rain (August through September) in millimeters}$$

$$TMP = \text{average temperature during growing season (April through September) in degrees centigrade}$$

$$Q = \text{a number Ashenfelter calls the quality index, defined in terms of the auction process for about 80 bottles of wine after they have matured in the bottle}$$

Rewrite the Bordeaux Equation in function notation. How many independent variables are there? What are they?

You have learned to distinguish between linear and nonlinear functions when a single variable is involved. The same concept can be extended to functions of more than one variable.

A function having the form

$$f(x_1, x_2, \ldots, x_n) = a_0 + a_1 x_1 + a_2 x_2 + \cdots + a_n x_n$$

where the a's are constants and the x's are the independent variables, is called a linear function of n variables. (Notice that no variable is multiplying or dividing another and that each variable is raised to the first power only.)

Refer to the functions you wrote for the preparation section. Compare with your partners to be sure you are in agreement. Decide which of the functions are linear.

---

[1] Copyright ©1990 by The New York Times Company. Reprinted by permission.

Examine the functions from Labs 1 and 2:

$$F = \frac{9}{5}C + 32$$

and

$$d = 16t^2$$

Is either one linear? What does its graph look like? How can you tell from the **formula** that one isn't linear? How can you tell from its **graph**?

Now, back to the Bordeaux Equation: In order to determine Q for a given year, how many inputs are required? Is Q a **linear** function of those variables or not? Which do you think would contribute more to the quality of a particular vintage: a slight increase in the average temperature during the growing season or a slight increase in the amount of winter rain? Which would you say is more desirable to the wine grower: winter rain or harvest rain? How did you determine your answers to the last two questions?

In 1988, the winter rains were about average, the temperature during the growing season was above average, and August and September were unusually dry. Since the flavor of the wine does not develop fully in the bottle for at least ten years, a definitive testing cannot take place before 1998. If Orley Ashenfelter had the opportunity now to buy the 1988 Bordeaux cheaply, do you think he would? Justify your answer.

Suppose in 1990 the winter rain and average temperature are about identical to those in 1989, but September 1990 is a very rainy month. Does Q increase or decrease with the additional September rainfall?

Let's consider the difficulties inherent in drawing graphs for a function of more than one input variable.

Topographical maps are one solution. They plot altitude as a function of both latitude and longitude by means of **level curves** for different elevations, while the input variables monopolize the horizontal and vertical axes.

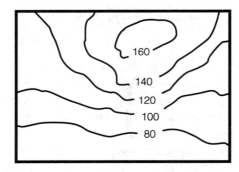

The same information could have been presented using different shades of color to indicate different elevations. You have seen examples of this kind of graphical presentation in an atlas. Even so, these solutions take care of only **two** independent variables. With three or more variables, we need to be either extremely creative (and artistically gifted), or able to take an easier way out.

Perhaps the most straightforward method is to pretend that all of the variables except one are constant, and then study the effects that changes in just that variable have on the function. In other words, we convert the multivariable problem to a single variable problem.

Suppose that in 1991 WR is 583 millimeters. Use your graphing utility to examine how Q varies with HR for

$$TMP = 15$$

$$TMP = 17.5$$

$$TMP = 20$$

Sketch these three graphs on the same set of axes. Be sure to label the axes and identify each graph. Be sure, also, that your sketch includes whatever portion of the graph makes sense in context (not much else is necessary). (Consider the units in which HR is measured. How large is a millimeter?)

Keep in mind always that the purpose of a graph is to convey information. Does your graph give a visual image of how Q depends upon HR for various temperatures? Should higher values of TMP be considered? Recall what temperatures TMP represents and discuss whether you ought to include any more graphs in your sketch.

Then, suppose $TMP = 16$ degrees. Use your graphing utility to examine how Q varies with WR for

$$HR = 50$$

$$HR = 170$$

Sketch these two graphs on another set of axes, again making sure that you show a sensible portion of the graph. Would a person examining your graphs understand what makes the quality of wine go up and what makes it go down? Recall that the range of a function is the set of output values. Do you think the range of this function could include negative numbers? Explain.

In Lab 1, we talked about determining different scales to measure temperature. In this problem, we are using a scale to predict the quality of wine. Most of the numbers generated by the Bordeaux Equation come in a cluster between about 3 and 4.6. To spread out these values of Q, we consider them as *exponents*. Their base is the irrational number e, which we will discuss in Lab 9. In Lab 11, we will see how to handle data with the opposite problem: the numbers are spread over a wide range of magnitudes. We will see that, in that case, we compare the exponents of the numbers. For now, all you need to know is that the constant -12.145 in the equation was chosen so that, when the variables representing winter rain, average temperature, and harvest rain were replaced by their values for 1961, Q turned out to be 4.6052. Since $e^{4.6052} = 100$ (check this on your calculator), 1961 is ranked by Ashenfelter as 100. (1961 was a spectacular vintage, and we are accustomed to a ranking system where 100 is "perfect.")

The Bordeaux ratings given below are all predictions, since wine must mature for at least ten years for its quality to develop fully. In the table, the results of Orley Ashenfelter's calculations for Médoc and Graves wines are compared with the ratings of two prominent wine critics.

Year	Ashenfelter's rating (1961=100)	Hugh Johnson's rating (scale of 0 to 10)	Robert M. Parker, Jr.'s rating
1987	38	3 to 6	pleasant, soft, clean, fruity
1986	23	6 to 9	very good, sometimes exceptional
1985	65	6 to 8	soft, fragrant, very good
1984	33	4 to 7	austere, mediocre quality
1983	76	6 to 9	superior to 1981, rarely achieves greatness
1982	56	8 to 10	most complex and interesting wines since 1961
1981	42	5 to 8	lacks generosity and richness
1980	28	4 to 7	light and disappointing

Copyright © 1990 by The New York Times Company. Reprinted by permission.

Ashenfelter's calculations, based on his model, suggest that 1989 Bordeaux wines will be the greatest of the century. Would you be willing to invest in wine futures for the 1989 Bordeaux vintages? That is, do you think the wine equation is a good model for determining the quality of a particular vintage? Support your answer with references to the table.

THE LABORATORY REPORT

Functions requiring two or more inputs are very common in areas from weather forecasting to economics. (These functions are called **multivariable functions**). Give three examples of functions that are used as models and that require two or more independent variables. State what the variables measure. Include at least one linear function.

Explain how you have learned to distinguish linear functions from nonlinear ones when there is more than one input variable. Discuss the difficulties of graphing a multivariable function and describe the method you used to produce graphs for the Bordeaux Equation. Include, as illustrations, the graphs you drew.

Explain how you were able to determine whether each variable had a favorable or unfavorable effect on the quality of the wine, and explain how you could tell whether one variable had a stronger effect than another. Give your group's evaluation of the Bordeaux Equation as a model for predicting wine quality.

Homework 5.1: A Boring Problem

1. $F(u, v) = 2u - 3v$ is a linear function of two independent variables.

(a) Holding v constant, first at -2, then at 0, then at 2, sketch three graphs of $F(u, v)$, showing how F varies with u. Label the axes (the variables in this case are u and F) and identify each graph. Put a scale on each axis.

(b) Holding u constant, first at 3, then at 1, then at -1, sketch three more graphs of $F(u, v)$, showing how F varies with v. Label the axes and identify each graph. Provide scales.

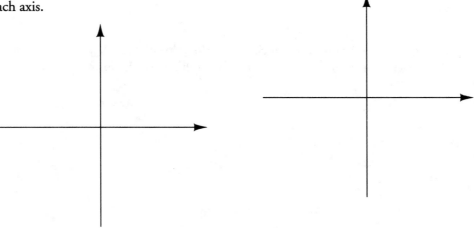

(c) Does an increase in the value of v cause the value of F to increase or decrease? How can you tell that from the formula?

(d) Which would cause a bigger change in the value of F: a change of one unit in u or a change in one unit of v? How can you tell that from the formula?

Homework 5.2: Quilts

These are three traditional American quilt patterns:

Mohawk Trail   Friendship Star   Drunkard's Path

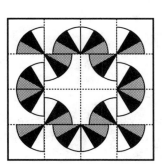 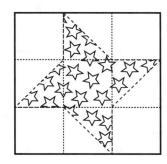 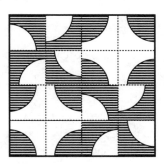

1. A quilter decides to incorporate 10-inch squares from all three patterns in her quilt, together with some plain square patches. These last are obviously the easiest to make; in fact she can make eight plain patches in an hour. By contrast, she can only make two Friendship Star squares in an hour. In hours per square, what is her rate for the plain squares? What is her rate for the Friendship Star squares?

   In the Drunkard's Path pattern, a curved piece is cut out of a small dark $2\frac{1}{2}$-inch square and replaced with an identical light one. This pattern is much more work because the curves are tricky to sew; it takes the quilter 3 hours to make a 10-inch Drunkard's Path square. The Mohawk Trail pattern is even more difficult, since each curved portion is composed of three separate pieces. It takes the seamstress 5 hours to make each such 10-inch square.

THE QUILTING FUNCTION

The length of time that it takes the quilter to complete a quilt depends upon the size of the quilt she makes and how many squares of each type she uses.

Let s be the number of Friendship Star squares in the quilt.

Let d be the number of Drunkard's Path squares in the quilt.

Let m be the number of Mohawk Trail squares in the quilt.

Let p be the number of plain squares in the quilt.

2. Using the rates you computed above, write a function of these four variables, $t(s, d, m, p)$, that gives the time required to construct the quilt in terms of the numbers of squares of the different types. Is your function linear? Explain your answer in terms of the formula for the function.

3. Suppose the sewer knows she wants a 60-inch by 80-inch quilt for a queen-size bed. What is the total number of squares she needs?

   Can you use this information to restrict the allowable values for the independent variables s, d, m, p? Write any restrictions that the quilting problem imposes on the domain variables s, d, m, and p.

4. Forget about time for the moment. Since you know the total number of squares in the quilt, you can write an equation for the sum of all the variables, and then solve for any one of them. Try solving for d. Now show how you could use your new equation to write t as a function of just the three variables p, s, m.

   Is your new function linear? Explain your answer in terms of the formula of the function.

5. Suppose the quilter decides the Mohawk Trail squares are so difficult that she will use only four at the very center of her quilt. Use your graphing utility to see how the dependent variable t varies with p for

$$s = 10$$

$$s = 15$$

$$s = 20$$

Sketch these three graphs on the same set of axes, labeling the axes and identifying each graph. Use your graphs to complete the assignment.

Discuss similarities and differences between the graphs.

6. As p increases, what happens to t? Explain why this makes sense in the context of the problem.

7. For a given value of p, which of your three graphs has the smallest t-value? (This means you need to fix the value of p and compare the three graphs with one another at that p-value.) Explain why this makes sense in the context of the quilting problem. (This requires a little hard thinking!)

8. Recall that the range of a function is the set of output values. Estimate the range of t if we restrict the quilter to 10-inch squares of the 4 patterns discussed above and a 60-inch by 80-inch quilt, but impose no other restrictions.

Laboratory 6

PREPARATION

Before meeting with your group, read through the lab sheets to get an understanding of the problem as a whole. Do the volume calculations for the test-case packages (two rectangular boxes and one cylinder). Work on the volume formulas and try to come up with functions $V(X)$ for the rectangular box and $V(r)$ for the cylinder.

At the beginning of the lab period, you should compare the numbers and the formulas you got with those of your partners and iron out any discrepancies. You should also decide in advance what information you will need to get during the lab from the graphing utility.

LABORATORY 6: PACKAGES

"Signed, sealed, delivered . . . "

The following note appeared in the June 27, 1991 *Postal Bulletin*, a publication of the U.S. Postal Service—one more example of the primal urge to standardize!

LENGTH OF UNIFORM GARMENTS/CARRIER AND MOTOR VEHICLE SOCKS AND HOSE

The following clarifications respond to field inquiries about certain provisions of the Uniform and Work Clothes Program. A future revision of the *Employee and Labor Relations Manual* will include these revisions.

Length of Shorts, Culottes, Skirts, Jumpers

Garments should not be more than 3 inches above mid-knee. Employees should not alter the length of their garments, and vendors are not authorized to make alterations that have hems falling more than 3 inches above mid-knee.

However, since everyone is not the perfect ratio of height to girth, some alterations may be necessary. Common sense must prevail in some situations. For example: If an individual stands 6 foot, 4 inches tall with a 34-inch waist, it is likely that the hem of his/her garment will fall more than 3 inches above mid-knee even when unaltered. If left as originally manufactured, the garment should still have a reasonable appearance.

Carrier/Motor Vehicle Socks and Hose

Currently, only black knee-length socks are authorized with walking shorts. However, specific requirements for skirts and culottes do not exist. Although these apparel items are an option for female employees, the National Uniform Committee intended that employees wear either the black knee-length hose, neutral-colored hose, or a coordinated, colored sock. Bright, fluorescent hose and socks are not permitted.

New socks have been designed to complement the carrier and motor vehicle uniforms and will be available for purchase soon. The *Postal Bulletin* will announce their availability.

—Labor Relations Dept., 6-27-91

The United States Postal Service will accept as fourth-class domestic parcels objects of a variety of shapes and sizes, provided that the weight does not exceed 70 pounds and that the length added to the girth of the parcel does not exceed 108 inches. The length is defined to be the measurement of whatever is the longest side of the parcel. The girth is the distance around the parcel at its thickest part.

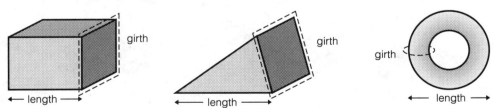

Let us suppose that we have a substance to mail, that it can be packaged in any shape whatsoever, and that we'd like to be able to ship as much of it as possible in a single package. (We will assume that the substance is sufficiently lightweight so that we don't have to worry about the 70 lb restriction.)

First, we will consider a rectangular box, and we'll suppose that the cross section of the box is square. In the lab "Galileo," we learned that, for a given perimeter, a square yields more area than any other rectangle, so it seems reasonable that we'd want to make the cross section of this box a square.

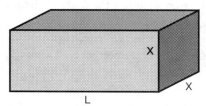

To check out this assumption, compute the volume for a couple of specific boxes. Let the length be 44 inches, and let the cross section be a 12-by-20-inch rectangle. Verify that length plus girth equals 108 inches. What is the volume? Now, keep the length at 44, but make the cross section a 16-inch square. Do you get more volume? This, of course, doesn't prove that a square is best, but it does at least support our intuition that a "flattened" box wouldn't hold as much as a "boxy" one. (If you laid a sealed half-gallon carton of milk on its side and jumped on it, you'd have a mess, because the flattened carton can't hold as much, even though its girth hasn't changed.)

Write a formula for the volume of the box in terms of L and X. Then use the fact that the length plus the girth equals 108 inches (assume we'll use up the allowable 108 inches) to express the relationship between L and X.

In order to optimize V (that is, find the greatest volume), we need to express V as a function of X alone. Use the relationship that you just wrote to assist you in writing a formula for V that has no variable other than X.

What type of function is $V(X)$? Use your graphing utility to draw its graph. Be sure to choose a viewing window large enough to see its important features: any X-intercepts, any turning points, and how it behaves beyond the intercepts. You may need to experiment with viewing windows and changes of scale before you get a meaningful picture. Is there a maximum or minimum value for $V(X)$?

You should observe that there is no maximum value for $V(X)$ because the function increases without bound as X moves away from the origin to the left. Similarly, there's no minimum value for $V(X)$. Why not? However, the graph has two turning points, one a **local minimum** and the other a **local maximum**.

Now consider only the portion of the graph that makes sense in the problem situation. (Recall the lab "Galileo," in which you used an abstract parabola to model a physical situation, and where you needed to trace the part of the graph that represented a falling body.) What is the smallest value of X that makes sense? What is the largest? Why? Look at the portion of the graph of $V(X)$ on that X-interval. Is there a local maximum? Use the trace feature of your graphing utility to locate the highest point. What is the value of X at that point? What is V? These numbers tell you something about the package. How long should it be? What is its girth?

One interesting distinction to make here is that between the shape of the box and the shape of the function representing the volume of the box. The curve $V(X)$ on your screen certainly does not suggest the shape of the mailing container, yet it expresses information about the volume of the container. What would be the volume of the box if $X = 0$? if $X = 10$? if $X = 20$? if $X = 27$? Your lab report should explain how the curve can represent the volume of the box as the dimensions of the box vary.

In Lab 2, "Galileo," we saw that a circle whose circumference is equal to the perimeter of a square has a larger area than the square, so it seems reasonable that a cylindrical container satisfying postal regulations might hold more than a rectangular box. Try $L = 44$ inches and see what the volume would be. (Notice that here, the girth is the circumference, so you would have 64 inches for the circumference. Also, to find the volume, you need to know the radius. How can you find the radius if you know the circumference?) Compare the volume with the volumes you already calculated for the two rectangular boxes of the same length. (It should be larger.)

Let's try to find the best cylindrical container. As before, you will need to write a formula for the volume and then rewrite it in terms of a single variable. First write V in terms of L and r. Then use the fact that length plus girth will equal 108 inches in order to rewrite V as a function of r. (You'll need to write the girth as a function of r.)

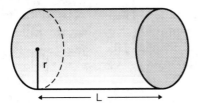

What type of function is $V(r)$? Does it have a maximum or minimum value? Consider only the portion of the graph that makes sense in this context: on what interval can r represent the radius of the cylinder? Look at the graph on that interval. You should see a local maximum. Locate it and explain what the coordinates mean in terms of the package. How long should the package be? What will be its girth? its radius?

Compare the volumes of the optimal cylindrical parcel and the optimal rectangular box. Is a circular cross section a more efficient use of girth than a square cross section? All other things being equal, which shape would you choose? Why?

Perhaps it would be advantageous to make the package round in every dimension. The Postal Service permits the mailing of spherical packages, provided they meet the weight and dimension restrictions. (Wrapping the package is, of course, a separate problem!) Perform the same analysis for a sphere and determine the maximum volume you would be permitted to mail in a single package. Of the three shapes, which would you choose?

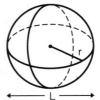

THE LABORATORY REPORT

Your report should show how you used polynomial functions to model the volume of a container, even though the shape of the graph did not suggest the shape of the container. You should explain the difference between the entire polynomial function (and its graph) and the portion that represents the problem situation. You should tell how you used each graph to determine the dimensions that would give the maximum volume for a particular shape, and give the maximum volume and best dimensions in each case. Discuss the relative merits of different shapes, under the particular restrictions of the Postal Service, and explain in everyday terms why you think one shape yields more volume than the others.

Homework 6.1: Exploring Polynomial Graphs

Cubics

In Lab 6, you used two specific cubic polynomials to model particular physical situations. In this assignment, you will examine the behavior of a variety of cubic polynomials.

The choice of an interval over which to graph a function and the relative scales used on the x- and y-axes can make a huge difference in how well you can see some of the features you'll be looking for. Be aware that the graphing utility may distort the apparent shape of a graph in order to fit it on the screen. You can minimize the distortion by changing the interval. Experiment.

1. Recall the shape of the graph of the simplest cubic, x^3. You have already learned what happens to that shape if you make slight variations, such as $2x^3$ or $-(x+1)^3$. Make a quick sketch of those three graphs. (You don't need a graphing utility for this.) How many x-intercepts does each one have? Be sure your sketch shows precisely where the x-intercepts are.

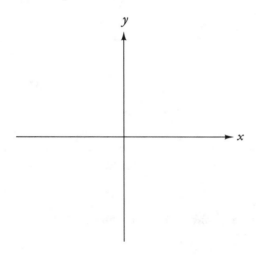

2. Now try graphing the following cubic. Use your graphing utility, of course.

$$f(x) = x^3 - 6x^2 - 11x + 6$$

Use a narrow interval on the x-axis. You'll be interested mostly in knowing how many times and approximately where the graph crosses the x-axis. You will probably have to play around with the interval before you feel satisfied with your picture. Try using the same scale on both the x- and the y-axes, so you can picture the true steepness of the curve.

3. Sketch this curve, indicating the approximate location of the x-intercept(s). Does this graph have anything in common with the graph of x^3 ? What features remain the same? What is different?

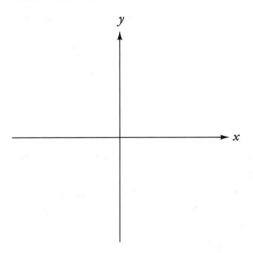

4. Clear the screen and try two more, one at a time:

$$g(x) = x^3 - 5x^2 + 8x - 4$$

and

$$h(x) = x^3 + x^2 + x - 3$$

You may need to change the scaling of the y-axis and/or the choice of interval. Features to note include these:

- number of roots (x-intercepts)

- number of turning points (local maxima or minima)

- whether the ends of the graph go up or down.

(A turning point isn't simply a wiggle in the graph, as you've seen in the graph of x^3, but a place where the graph changes from rising to falling, or vice versa.)

Note your observations:

5. Now graph these cubic polynomials, one at a time:

(a) $-x^3 - 6x^2 + 11x - 6$

(b) $-x^3 - 5x^2 + 8x + 4$

(c) $-x^3 + x^2 + 3x + 4$

(d) $-x^3 - x^2 - 3x + 4$

Observe similarities and differences and note your observations.

Try some more cubics, including some whose leading coefficient (the constant multiplying the x^3) is something other than 1 or -1.

6. At this point, you should be ready to generalize about the graphs of polynomials of the form $a_3x^3 + a_2x^2 + a_1x + a_0$, where the a's are constants and a_3 doesn't equal zero.

(a) How many roots do they have? (Don't just say "three!" Not all cubics have three roots.) Tell all the possibilities.

(b) How many turning points are there?

(c) In which direction will the ends of the graph go, and how can you tell from the formula?

(d) Please illustrate your answers to (a) through (c) with some representative sketches.

Homework 6.2: Exploring Polynomial Graphs

General

In the cubic polynomial assignment, you became familiar with the possibilities for polynomial functions of degree three. In this assignment, you will explore the behavior of polynomial functions of other degrees.

1. You're already familiar with second-degree polynomials. What do their graphs look like?

2. What possibilities exist for the roots of a second-degree polynomial? Supply the commentary for these illustrations:

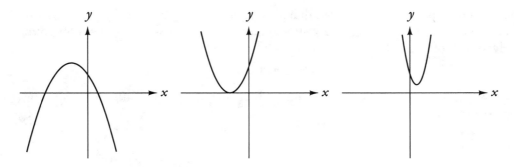

3. What's another name for a first-degree polynomial function? How many *x*-intercepts does its graph have?

4. What does the graph of a polynomial function of degree zero look like?

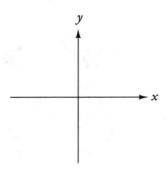

5. The graphs of polynomial functions of higher degree are harder to classify, but we can still make some generalizations. Using your graphing utility, investigate the graphs of several polynomials of degree four, such as

$$x^4 \qquad \text{and} \qquad -2x^4 + x^3 + 5x^2 - 2x + 3$$

Try changing just one constant at a time and see how the graph is affected. Note any observations you make.

6. Write the formula for a fourth-degree polynomial with no roots and sketch its graph.

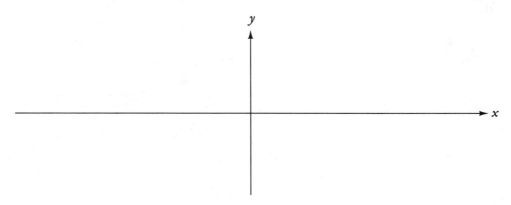

7. Find a fourth-degree polynomial with four roots. (**Hint:** you might decide first what you want the roots to be, and then write the polynomial in factored form.)

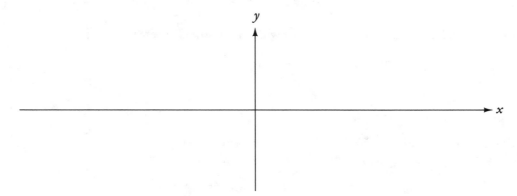

8. Find a fourth-degree polynomial with exactly two roots. Sketch its graph and give its formula, explaining how you found the formula.

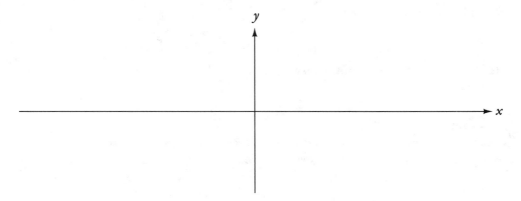

9. Sketch the graph and give the formula for a fourth-degree polynomial with exactly three (distinct) roots.

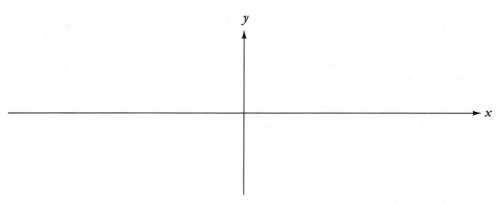

10. Can you find a fourth-degree polynomial with five roots? Give a reason for your answer.

11. From your explorations, you should now be ready to make some generalizations about the graphs of degree-four polynomials. Given some function of the form

$$f(x) = a_4 x^4 + a_3 x^3 + a_2 x^2 + a_1 x + a_0$$

where the a's are constants and a_4 is not equal to 0, answer the following questions.

(a) How many roots might there be? (Tell all the possibilities.)

(b) How many turning points might there be?

(c) How can you tell **from the formula** in which direction the ends of the graph will go?

Extend your explorations to polynomials of degree greater than four, with a view to answering questions (a) through (c) above.

12. Summarize what you have learned about the roots of a polynomial function.

   (a) What's the smallest number of roots an odd-degree polynomial can have?

   (b) What's the smallest number of roots an even-degree polynomial can have?

   (c) What's the connection between the greatest number of roots a polynomial can have and the degree of the polynomial?

13. Summarize what you have learned about turning points. What is the maximum number of turning points the graph of a polynomial function can have? (How is that number related to the degree of the polynomial?)

14. If you know that a polynomial function has degree 5 or 7 (or some other odd number), what can you say about the ends of the graph?

15. If you know that a polynomial has even degree, what can you say about the ends of the graph?

Homework 6.3: Optimize for a Ban on Waste

In recent years, people concerned about the environment have campaigned to encourage moderation and conservation in our use of resources. The U.S. Public Interest Research Group (PIRG) presents "Wastemaker Awards" to manufacturers who it claims use unnecessary amounts of packaging for some of their products. Health and beauty aids appear to be the worst offenders, winning all seven of the 1991 awards.

One of these dubious honors went to Bristol-Myers Products for its Ban Roll-On deodorant. One-and-a-half fluid ounces (44 cubic centimeters) of liquid come packaged in a plastic container, mainly cylindrical in shape but pinched in the middle. The cylinder is 5.1 cm high and has a diameter of 3.8 cm. It is topped by a plastic cap 4.1 cm high, having the same diameter. All of this rests in an outer container of cardboard, a rectangular box whose dimensions are 13.3 cm by 7.0 cm by 4.1 cm. There is, in addition, some inner cushioning, also made of cardboard.

Calculate the approximate surface area of the packaging materials used for a 1.5 ounce container of Ban Roll-On. How much is cardboard (you may ignore the inner cushioning as well as any overlapping of flaps that would be found in a box)? How much is plastic (again, you may ignore the region where the cap overlaps the main container)? The plastic would be tricky, because the shape isn't a perfect cylinder, so you'll have to pretend that it is one. Will your answer be an overestimate or an underestimate of the actual amount of plastic used? Explain.

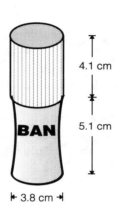

1. Now you will figure out just how efficient Bristol-Myers might have been with their plastic container. Suppose you wanted to package 44 cubic centimeters of liquid in a cylindrical container. (We're going to ignore details such as providing a way to open the package.) Write a function that expresses the surface area S as a function of r and h. Remember to include the top and the bottom.

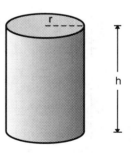

Express the volume V as a function of r and h.

2. Since you have determined that V is to be 44 cubic centimeters, use 44 for V. Now you can solve for one of the variables in terms of the other. Substitute that expression into your surface-area function. The surface-area function should now be a function of a single variable. What is the independent variable?

3. Your goal is to minimize the surface area, and thus the amount of plastic. Can you tell from the formula whether the function has a minimum? Why or why not?

4. Use your graphing utility to graph the function. Choose a sensible interval for the independent variable and look for a minimum in that interval. (You should see a curve that dips down and then rises. If your graph doesn't have the right shape, check to see whether you've enclosed the denominator in parentheses.) From the graph, determine the best value for the independent variable. Make a sketch of the relevant portion of the graph, indicating the interval you have used and labeling the coordinates of the minimum.

5. Give the dimensions of the most efficient cylinder.

    (a) How much plastic does it use (surface area)?

    (b) What is the ratio of r to h?

6. Conserving materials doesn't appear to be one of the major concerns of this manufacturer. In designing the package, what might be some other considerations? Do you think your optimal cylinder would serve its purpose well? Do you think the cardboard is necessary or important? Give reasons.

Laboratory 7

PREPARATION

Prepare for this project by reading the lab sheets and writing formulas for the three average-cost functions F, G, and H. Bring the formulas with you to the lab, and begin by comparing your functions with those of your partners.

$$F(x) = \rule{4in}{0.4pt}$$

$$G(x) = \rule{4in}{0.4pt}$$

$$H(x) = \rule{4in}{0.4pt}$$

LABORATORY 7: DOORMATS

"These boots were made for walking . . . "   —*Nancy Sinatra*

"I myself have never been able to find out precisely what feminism is: I only know that people
call me a feminist whenever I express sentiments that differentiate me from a doormat."
—*Rebecca West, 1913*

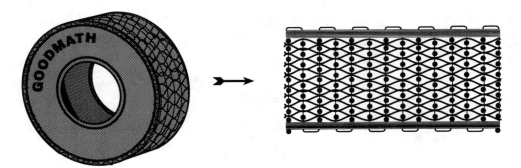

A western Massachusetts couple runs a small cottage industry from their home. They
recycle discarded automobile tires by slicing them up into strips and making doormats,
somewhere between five hundred and two thousand doormats each year. Their somewhat
imprecise method of cutting the tires results in a product of irregular thickness, perfect
for scraping mud and snow off shoes. Worn-out tires are a disposal problem nationwide,
so they have no difficulty acquiring their raw materials. Additional materials are few and
inexpensive—some colored plastic beads for spacing and some thick aluminum wire for
holding everything together.

The couple has figured out that, counting the cost of the raw materials and the value
of their labor, it costs them $5 to produce each doormat. In addition, they estimate that
their business has fixed annual expenses of $200. (This is a very "low-tech" operation.)
In the language of economics, the capital cost is $200 and the variable cost is $5 per unit.
They want to determine the average cost of each mat they make this year.

THE SIMPLEST MODEL

First write a formula for the total cost of producing x doormats this year. Now write a
function that expresses the average cost of a doormat. Using the graphing utility, explore
the behavior of this function, which we'll call F. Is the average cost a constant? If so, what
constant? If not, why not?

What's the average cost if they produce 200 mats this year? 500? What happens to
the average cost if they produce very few?

As the number of doormats produced increases, what happens to the average cost?
(Economists call this effect "spreading the overhead.") Is it reasonable to suppose that, as
more and more doormats are made, the average cost would continue to drop? If you study
the graph for large values of x, you should see that the curve appears to level off. Is there a
lowest possible average cost? Verify that, although the values of the function get smaller and
smaller, there's a limit below which the average cost cannot go. From the graph, determine

this limit. Does the function ever attain that value? In other words, if we call the limit L, can you find a value of x such that $F(x) = L$? Give a reason for your answer.

Enlarge the x-interval of the viewing window so that it's at least $10,000$ units wide. You may include negative values of x if you wish. The graph of $F(x)$, except near the y-axis, should resemble a horizontal line. Give the equation of that horizontal line. Because the graph of $F(x)$, in the long run, resembles the line, we call that line the **horizontal asymptote** for the function F.

The function F as you've written it probably has the form $F(x) = \frac{a + bx}{x}$. Rewrite it (by dividing) in the form $F(x) = \frac{a}{x} + b$.

If the magnitude of x is very large, what can you say about the value of $\frac{a}{x}$? Explain how your answer to that question shows that, for large values of x, the graph of $F(x)$ resembles the line $y = b$. We say that the graph of $F(x)$ has the horizontal line $y = b$ as an asymptote, that the function approaches the value b asymptotically.

A MODEL THAT ACCOUNTS FOR EFFICIENCY

The mat makers have discovered that they become more efficient economically if they make larger numbers of mats. Perhaps they can get a better price on the beads, or perhaps their enthusiasm propels them to work more quickly. You can express this mathematically by including a quadratic term in the total cost function. Suppose that term has been determined to be $.0003x^2$.

Rewrite the total cost function, including the quadratic term (you will need to decide whether to add or subtract this term). Write a new average-cost function; call it G.

Use the graphing utility to compare the behaviors of $F(x)$ and $G(x)$ over several different intervals, using a variety of viewing windows. Find intervals over which they look nearly the same. Find out approximately where they begin to diverge. At what number of mats does the x^2 term appear to "kick in" and become significant? (This is a matter of opinion; your answer will depend upon how you define *significant*.) Look at the formulas for the two functions and explain why the graphs are very close for certain values of x.

If you use sufficiently large values of x, you should observe that the graph of $G(x)$ begins to look like a straight line rather than a curve. This line, though, should appear slanted rather than horizontal. It is called an *oblique*, or *slant*, asymptote for the function G. Here are two methods for discovering an equation for that line.

First method

Using an very large interval, centered at 0, such as $-10,000 < x < 10,000$, examine the graph of $G(x)$. Except for some bizarre behavior near the y-axis, it should resemble a straight line. Since you can write an equation for a line if you know two points, use the trace feature to find the coordinates of a point far to the left on the screen and of another one far to the right. Write the equation in $y = mx + b$ form.

Second method

As you did in "The Simplest Model," rewrite the formula for $G(x)$ by dividing, so that is now in the form $\frac{a}{x} + b + cx$. Think about what happens to each term as x grows very large. Compare the results with those you obtained with the first method and explain how you can use the formula for the function to determine its asymptote.

What does this have to do with making doormats? Go back to some realistic values for x and interpret the behavior of the graph in the context of this problem.

A MODEL THAT INTRODUCES SOME INEFFICIENCY

The mat makers have also observed that, if they spend all of their energies making doormats, their productivity declines, and therefore the cost to them begins to rise again. You can express this mathematically by including a cubic term in the total cost function. Decide whether the term ought to be added or subtracted.

Suppose the cubic term has been determined to be $.0000001x^3$. Rewrite the total cost function to include this term. (Hang onto the quadratic term as well.) Find a third average-cost function, $H(x)$, which includes the cubic term. Then use the graphing utility to examine all three functions over an interval that makes sense in context. Does the graph of $H(x)$ suggest an optimal number of doormats for the doormat makers to produce?

Now investigate the long-run behavior of $H(x)$. Enlarge the viewing window so that the dramatic behavior near the y-axis is minimized. You'll probably need an x-interval of $40,000$ units or so. In this large viewing window, what is the apparent shape of the graph of $H(x)$?

Check this out algebraically: rewrite the formula for $H(x)$ by dividing, as you did in the last two sections. How many terms are there? Which one is of negligible size whenever x is a huge number? Explain why, for enormous values of x, the graph of $H(x)$ looks the way it does.

The function H has an asymptote that is neither horizontal nor oblique; its asymptote is another curve. Give the formula for that curve. Overlay the graph of the curved asymptote, try several intervals, and observe that the graph of $H(x)$, from a distance, resembles the graph of its asymptote.

So, what about doormats? Interpret the asymptotic behavior of the function H in the context of this cottage industry.

THE LABORATORY REPORT

You are serving as a mathematics consultant to the doormat makers. Tell them about the three different functions used to model average cost. Explain what each term in the function represents and why the term is positive or negative. Compare and contrast the functions for their ability to model a concrete situation.

Describe the long-term behavior of each of the functions. (Sketches would be useful here.) Show how you used algebra to predict the asymptote for each function.

Conclude with three separate recommendations for the mat makers. What might they want to adopt as production goals if they were to use model F? What advice would you give them if they decided that G were a more accurate model for their business? And finally, how many mats would you recommend they produce if model H appeared to be the best fit for their operation?

Homework 7.1: Black Holes and Vertical Asymptotes

"Il est donc possible que les plus grands corps . . . de l'univers, soient . . . invisibles."
—*Pierre Simon, Marquis de Laplace*

[It is therefore possible that the largest bodies . . . of the universe, are . . . invisible.]
—*Author's translation*

In Lab 7, you explored what happens to a rational function in the long run, that is, when the magnitude of the independent variable gets very large. In this assignment, you will investigate the behavior of the graph of a rational function in the vicinity of a number that isn't in the domain of the function.

1. Using the graphing utility, draw several graphs of functions having the form

$$f(x) = \frac{x^2 - k}{x - 3}$$

(You choose values for k.) Experiment with the viewing window until you are able to see two distinct branches for each of the graphs (or for all but one of the graphs). Your best viewing window will probably be one that's only about six or eight units wide.

2. Notice that 3 is not in the domain of any of these functions. Arrange the viewing window so that $x = 3$ is at the center and observe the dramatic behavior of the curves as they approach $x = 3$. You should see that they become nearly vertical. Why? What is happening to the denominator of each function as the value of x approaches 3? Explain how those denominators affect the functions and their graphs.

The line $x = 3$ is called a **vertical asymptote** for those functions. The closer x gets to 3, the more the graphs actually resemble that vertical line.

3. You may have noticed that what was just stated is not entirely true! Was this one of your functions?

$$g(x) = \frac{x^2 - 9}{x - 3}$$

If not, graph it now.

What's going on here? Why does the graph of $g(x)$ look so different from all the others? Look at the algebraic formulas, and explain the difference between $g(x)$ and the other functions you graphed.

4. If you look closely at your graphing utility, you may be able to find a hole in the graph at $x = 3$. Even if you don't see any hole, you know that the function isn't defined for $x = 3$ so, despite what the graphing utility seems to indicate, the graph actually stops on one side of 3 and starts up again on the other side. Except for the hole, to what simpler function is $g(x)$ equivalent?

5. Sketch a graph of $g(x)$. How will you indicate that the number 3 is not in its domain? Label the coordinates of the hole.

6. Find the formula for another rational function that has a hole. Sketch its graph, showing clearly where the hole occurs.

7. Sketch the graph of $f(x) = (x+2)/(x^2 + x - 2)$, showing clearly any vertical asymptotes or holes.

Homework 7.2: Long-term Behavior of Rational Functions

"From here to eternity"

In Lab 7, you studied the graphs of three rational functions. You saw that, for very large values of the independent variable, the graphs of those rational functions resemble the graphs of simpler functions—a line or a parabola. In this assignment, you will explore several rational functions and try to come up with a method of predicting their nonvertical asymptotes, that is, the simpler functions they eventually resemble.

A rational function is defined to be the quotient of two polynomial functions, where the polynomial functions can be as simple or as complicated as we wish. In Lab 7, the denominator for each of the rational functions was the very uncomplicated polynomial x, and it was relatively easy to figure out what those graphs were going to look like in the long run.

1. Define a function that has the form $(ax + b)/(cx + d)$, that is, a linear expression divided by another linear expression. (You choose values for a, b, c, and d.) What sort of asymptotic behavior might be expected? Use your graphing utility to investigate the graph of your function over a very wide viewing window. Does it level off? Does it resemble a curve? Use the graph to write an equation for the asymptote. (Choose two points, one from each edge of the screen.)

Your function	Its asymptote

2. Now get some help from algebra: using long division, divide the denominator into the numerator. You should get a **number** plus a remainder. In other words, you've shown that

$$\frac{ax + b}{cx + d} = n + \frac{m}{cx + d}$$

where m and n are other constants.

When x is very large, what can you say about the quantity $\frac{m}{cx+d}$? If you weren't sure about the asymptote before, you should be able to read its equation from the calculations you've just done.

3. Choose two other functions that have the same form (linear over linear). Predict their (nonvertical) asymptotes. Use the graphing utility to check.

Your functions	Their asymptotes

4. Try a function that's a quotient of two quadratic expressions, such as

$$\frac{2x^2 + 3x - 1}{3x^2 + 2}$$

Could you have predicted its asymptote?

5. Experiment with other forms of rational functions. Try one that has a linear expression on the top and a quadratic on the bottom. Do you have any expectations about the nonvertical asymptote?

   Try dividing the bottom into the top. You're stuck, right? Why? This has implications about the nonvertical asymptote. Graph the function and see what it looks like for large values of x. What is the asymptote?

Your function	Its asymptote

6. Look for a pattern by trying a couple more functions in which the bottom has a higher degree than the top. They all have the same horizontal asymptote. What is it?

Your functions	Their asymptotes

7. Now switch tops and bottoms, so that the top has a higher degree than the bottom. Try to figure out what's happening, both algebraically and graphically. For the graph, you need only look—in a very wide viewing window. For the algebra, you need to perform long division and think about what portion of the result is negligible when x becomes very large. Be sure that you're doing some examples that are other than a quadratic over a linear; you should include some expressions of higher degrees and you should include some functions in which there's at least a two-degree difference between the top and the bottom. (Aren't you glad you don't have to draw all these graphs by hand?)

Your functions	Their asymptotes

8. After enough exploration, you should be ready to draw some conclusions and write a summary of the long-term behavior of rational functions. There are really only three separate cases: (1) the degree of the numerator and the degree of the denominator are the same; (2) the degree of the numerator is less than the degree of the denominator; (3) the degree of the numerator is greater than the degree of the denominator. Tell what you learned about each case.

9. "Except near the vertical asymptotes, the graph of a rational function resembles that of a polynomial." (Recall that constant functions and linear functions are special cases of polynomials.) Tell how your work in this assignment would support that statement.

Laboratory 8

PREPARATION

"Hi, folks! This is Claire Sky, your fair-weather weather woman. The high temperature for today was 97° and the low for tonight will be around 80. It's been a scorcher! The cooling degree-days to date are topping 350. That's way above normal this early in the summer."

In addition to giving the high and low temperatures for the day, weather reports often include the number of heating or cooling degree-days for the season. We may tend to ignore this information, but if the number of heating degree-days for the winter is higher than normal, fuel bills can skyrocket. (Think about the impact on the fuel budget at your college, which in turn affects tuition costs. Think about the homeowner or renter who has to pay the electric bill for a house with electric heat.)

In Lab 4, you worked with models for power usage that depended on the average monthly temperature. However, the actual data supplied by Massachusetts Electric Company reported monthly heating and cooling degree-day totals. These monthly totals had to be converted into average monthly temperatures before the models for Lab 4 could be selected. Why would the electric company choose to store heating and cooling degree-day totals instead of simply monthly temperature totals (or averages)? How do you turn temperature readings into heating and cooling-degree days?

Start with a base temperature of 65°F, a comfortable temperature requiring little air conditioning or heating. This is the 0 degree-day level for both heating and cooling degree-days. If the average daily temperature is higher than 65°F (air conditioners will get turned on), the cooling degree-days for the day will be the difference between the average daily temperature and 65. The heating degree-days will be 0 (no heat needed). Suppose the average daily temperature is 76°F. What are the cooling degree-days? heating degree-days?

If the average daily temperature falls below 65°F (turn on the heat), the heating degree-days will be the difference between 65 and the average daily temperature. The cooling degree-days will be 0 (no air conditioning needed). How many heating degree-days will have accumulated from a day with an average temperature of 45°F? how many cooling degree-days?

Now sum up the heating and cooling degree-days for all the days in the month. That's the monthly temperature information that was collected by Massachusetts Electric Company. Examples from the data are presented below, along with the corresponding average monthly temperatures that were computed for Lab 4.

Year	Month	h	c	x
1986	Feb	1126	0	24.8
1987	July	32	13	64.4
1988	July	13	331	71.3
1989	June	113	87	64.1
1989	Oct	289	34	56.8

h = heating degree-day totals;

c = cooling degree-day totals;

x = average temperatures in °F

As the season changes from fall to winter, what do you expect to happen to monthly heating degree-days totals? How will power consumption be affected? The highest number of cooling degree-days typically occur in late summer. What effect does increasing the number of cooling degree-days have on power usage? Why?

Recall Model 2 from Lab 4, "Electric Power."

$$p_2(x) = 1953 - 28.7x + .25x^2$$

where x was the average monthly temperature and $p_2(x)$ was power consumption in millions of kilowatt hours.

Using $p_2(x)$, what would you predict for power usage during a month with an average monthly temperature of 64°F?

Consider the following scenarios for a month with average temperature of 64°F.

1. The average daily temperatures stayed fairly close to 64°F for the entire month.

2. The days in the beginning of the month were cold, but the end of the month turned out to be quite hot.

For which of the two situations do you think that power usage would be greater? Why? Does $p_2(x)$ discriminate between the two scenarios for the month with the 64° average? Do you think that $p_2(x)$ would tend to overestimate the power consumed in a month that behaved as described in scenario 1? If not, would $p_2(x)$ underestimate the power consumption?

Discuss your answers with your partners when you get to lab. Make sure all the members of your group understand how to compute heating and cooling degree-days from average daily temperature.

LABORATORY 8: MORE POWER

"Power to the people!"

As preparation for this lab, you discovered that the models for power consumption based on average monthly temperature cannot adjust for the situation in which two months have the same average temperatures, but their needs for air conditioning and heating are different. For example, the average temperatures for July 1987 and June 1989 are both about 64°F. But the heating (h) and cooling (c) degree-day totals are quite different.

	h	c
July 1987	32	13
June 1989	113	87

Describe what you think the temperature patterns for these two months were like. Would you predict higher power consumption for July 1987 or June 1989?

MODEL FOR POWER CONSUMPTION BASED ON HEATING AND COOLING DEGREE-DAYS

$$p(c, h) = 1000 + .309h + .853c$$

where $p(c, h)$ is the power consumption (in millions of kilowatt hours) corresponding to a month with a sum total of c cooling degree-days and h heating degree-days.

Use this model to estimate the power consumption for July 1987 and June 1989. How do these results fit in with what you predicted would happen?

During the winter months December, January, and February, the cooling degree-days have been 0 for all ten years that this data was collected, while the heating degree-days have ranged from around 760 to 1500. Graph the relationship between heating degree-days and power consumption that results from this model when $c = 0$. As the weather gets colder, what happens to the predicted power consumption?

June in New England often consists of a mixture of hot and cool weather. The average number of cooling degree-days for June was 82. The number of heating degree-days ranged from 85 to 181. Overlay the relationship between heating degree-days and power consumption when $c = 82$ and h is between 85 and 181. Now overlay the relationship for $c = 0$ and h is in the same range. How much effect did the cooling degree-days (the need for some air conditioning) have on the prediction for power consumption?

The number of heating degree-days for August is usually minimal. What will happen to the predicted power consumption from our model if we look at increasingly hot months for August?

Analysts employed at power companies must work to create models that are representative of patterns present in data sets. Let's focus on the idea of mathematical modeling and observe the effect of the addition of various terms to mathematical relationships. (The functions discussed below are not related to any actual data set. You may have to adjust the viewing window to get good pictures of what is occurring.)

Start with hypothetical Model A, a model that is linear in both h and c and thus is similar to our original model. Model B results when an *interaction* term $.01ch$ is added to Model A. Model C results when the *quadratic* term $.02h^2$ is added to Model A.

$$\text{Model A: } f_1(c, h) = 500 + 3h + 2.5c$$

$$\text{Model B: } f_2(c, h) = 500 + 3h + 2.5c + .01ch$$

$$\text{Model C: } f_3(c, h) = 500 + 3h + .02h^2 + 2.5c$$

First consider Model A. Hold c fixed at 100, then at 200, and then at 300. Overlay graphs of $f_1(c, h)$ for each of these fixed choices of c. What pattern appears? What is the change in power consumption associated with a one degree-day increase in h when c = 100? c = 200? c = 300? Does this change depend on the value fixed for c?

Repeat the work in the last paragraph for function $f_2(c, h)$. How does this model differ from Model A in terms of the graphs that you see? What happens to the rates of change in power consumption with respect to an increase of one heating degree-day when c is set at 100, then 200, and then 300?

Suppose you felt that in the real-life situation, the rate of change in power consumption with respect to heating degree-days was constant no matter what fixed value of c was chosen, but that this rate should be larger when c was 300 than when c was 100. Would you use the model that was linear in both c and h or would you use the model that included an interaction term? Explain your choice.

Now consider function $f_3(c, h)$. Hold c fixed at 100, then at 200, and finally at 300. Overlay graphs of $f_3(c, h)$ for each of these fixed choices. What pattern appears? Is the rate of change in power consumption with respect to heating degree-days the same for each of the three fixed values of c? Now reverse the roles of c and h. Hold h fixed at 100, then 200, and then 300. Again overlay the three graphs that result from Model C. Why do you get parabolas in the one case (c fixed) and lines in the other (h fixed)?

THE LABORATORY REPORT

Organize your answers to the questions contained in these lab sheets into your report. Discuss the added flexibility obtained by using heating and cooling degree-days in the model for power consumption over a model based solely on average monthly temperature (as in Lab 4). Graph and discuss patterns you observed while looking at the three hypothetical Models A, B, and C. What was the effect of adding the nonlinear terms to Model A?

Homework 8.1: "Lost in the Supermarket"

In manufacturing containers of various sizes and shapes, companies should understand the relationship between the dimensions they choose for the container and the amount of material that makes up the container. Consider, for example, an *open* box with a square bottom.

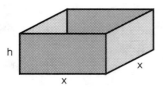

1. Let $S(x, h)$ be the surface area of the box. Write a formula for $S(x, h)$, the relationship between the amount of cardboard material in the box and the dimensions of the box.

The purpose of this problem is to analyze how the amount of material in the box is affected by changing the dimensions of the box.

POSSIBILITY 1: THE BASE OF THE BOX IS FIXED

2. Suppose the base of the box measures 1 inch by 1 inch, that is, $x = 1$. Rewrite your formula and graph the relationship between the amount of material and the height of the box, $S(1, h)$. How much additional cardboard will be required every time you increase the height of the box by 1 inch?

3. Now change the dimensions of the base and let $x = 2$. Graph $S(2, h)$. How much additional cardboard will be required every time you increase the height of the box by 1 inch? Repeat for $x = 3$.

Sketch the graphs $S(1, h)$, $S(2, h)$, $S(3, h)$ on a single axis.

You have used three different (fixed) values for x. When you increased the x-value, what happened to the **rate** of increase in surface area per 1-inch change in height?

POSSIBILITY 2: THE HEIGHT OF THE BOX IS FIXED

4. Suppose the height of the box is fixed at 1 inch. Graph the relationship between the amount of material and the base dimension $S(x, 1)$. How much more material will be required if you make a box that has a base of 3 inches instead of a base of 2 inches? a base of 4 inches instead of a base of 3 inches? a base of 5 inches instead of 4 inches?

5. Suppose you were to change the dimensions of the base of the box from $4\frac{1}{2}$ to $5\frac{1}{2}$ inches. Without doing the actual calculations, can you tell whether the amount of additional material you would need to make the box is more (or less) than you needed when you changed the dimensions of the base from 4 to 5 inches?

6. Overlay the graphs $S(x, 1)$, $S(x, 2)$, and $S(x, 3)$. Adjust your viewing screen so that $x \geq 0$. Sketch the graphs.

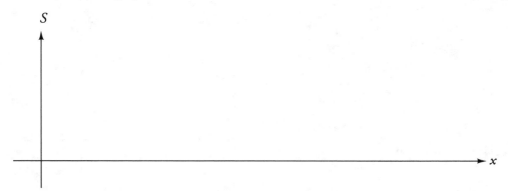

Which of the three graphs appears to be rising most steeply? Calculate the amount of additional material needed to make a box when the base is changed from 2 to 3 inches for a box of height $h = 1$; of height $h = 2$; of height $h = 3$.

7. Compare your two sets of graphs. What do they tell you about this hypothetical box?

Laboratory 9

PREPARATION

The 1980s saw a growing awareness and concern over the spread of AIDS. Apart from distress over the tragedies of individual cases was the realization on the part of public health officials that the mathematical model for the spread of the epidemic was an exponential function. Indeed, an essay written by the biologist Stephen Jay Gould for the *New York Times* (April 19, 1987) had the lead, "The exponential spread of AIDS underscores the tragedy of our delay in fighting one of nature's plagues." In the article, Gould states that "the AIDS pandemic . . . may rank with nuclear weaponry as the greatest danger of our era."

To appreciate the urgency of the problem, we need to understand the nature of exponential growth. The simplest example is a trick problem that many of you have seen in grammar school:

> If you place a penny on the first square of a checkerboard, two pennies on the second square, four coins on the third square, eight on the fourth, and continue doubling the number of pennies on each subsequent square, how many pennies are on just the last (64th) square?

Write the answer using an exponent, and then give the dollar value of the pennies. (**Hint:** if you could stack the pennies, you'd have a stack about as high as the universe is wide!)

Fill in the following table, in which the second column gives the number of pennies on the square whose number is in the first column.

Number of square	Number of pennies
1	
2	
3	
4	
5	
6	

Graph the points determined by the table. This is an example of *discrete* data; the only values that are meaningful for both the x- and the y-coordinates are positive integers.

To perceive the pattern of the data more clearly and to help determine the mathematical function we want to use to model the data, connect the points with a curve as smoothly as possible. Let y be the number of pennies on the xth square and write an equation giving y as a function of x. The function you determined is called an *exponential* function. Where does the independent variable occur in this function? _____

Make two more graphs, one for each data set given in the lab sheets. Bring all the graphs with you to the lab.

LABORATORY 9: AIDS

"Bridge over troubled water . . . "

Articles in the *New York Times* and *Daily Hampshire Gazette*, July 11, 1990, under the title "AIDS killing more young women," reported that HIV infection is rapidly becoming a major health problem in young women. The article gave this table of deaths of women aged 15–44 attributable to HIV/AIDS. Plot these points on a graph where the integers 0, 1, 2, . . . on the horizontal axis correspond to the years 1980, 1981, 1982, . . . and the numbers on the vertical axis correspond to the numbers of deaths. Approximate your data by a smooth curve and use your curve to estimate the number of deaths of women aged 15–44 due to AIDS in 1989.

Year	1980	1981	1982	1983	1984	1985	1986	1987	1988
Deaths	18	30	36	92	198	360	631	1016	1430

In the exponential function that we used to model the coins on a chessboard problem, the *base* was 2, since the number of coins doubled with every square. Compare with your partners the formulas each of you wrote for the chessboard problem. When you are in agreement, use your graphing utility to draw the function you determined. Experiment with viewing windows and changing scales on the axes until the graph on the screen matches the one you drew.

Actually, the most commonly used base in mathematics is the irrational number *e* ($\approx$ 2.7183). Using your graphing utility to help you, sketch the graphs of $f(x) = 2^x$, $g(x) = e^x$, $h(x) = 3^x$ in three different colors on the same set of axes, using a sheet of graph paper. Determine the domains and ranges of these abstract functions by looking at several different viewing windows. Then notice what features of the three graphs are the same and what are different. Notice the long-term behavior of these functions. Do the graphs have any asymptotes? (You should use your graphing utility to explore this question.) Experiment with your graphing utility to determine how a change in the base affects the graph of an exponential function. Try several bases less than 1. What happens when the base is negative?

Note: A function such as $y = -3^x$ does not have a negative base. The base of that function is 3; the function given is simply the *opposite* of the function $y = 3^x$. How would you have to write a function if it were to have a negative base?

Let's return to the AIDS statistics. In the real, noninfinite world, an exponential function can model a situation only temporarily. Since exponential functions increase so rapidly, they exhaust the population very quickly. (This is why chain letters are silly and ones that ask for money are illegal!) Let's try to model the data with a function of the form $y = c \cdot e^{kx}$, where c and k are constants that we wish to determine.

Using statistical techniques, it's possible to find a function of that form that fits the data pretty well:

$$y = 16e^{.586x}$$

Use your graphing utility to sketch the graph of the function and copy it onto the same set of axes on which you sketched the graph of the AIDS data. Use a color so you can distinguish the two curves. (**Warning:** when you type the function, you must enclose the entire exponent in parentheses, or the graphing utility may think that only the .586 is the exponent.)

For what years does your function model the data fairly well? How many deaths does the model give for 1989? Compare that number with the estimate you made from your hand-drawn graph. What is the number of deaths your model predicts for the year 1990?

There are ways to change the exponential nature of the growth of AIDS. An article in the January 5, 1990, *Los Angeles Times* titled "Slower Spread of AIDS in Gays Seen Nationally," says "Public health officials cite several possible causes for the slowdown, including the adoption of safer sexual practices by many gay men to prevent infection."

The following table gives the number of AIDS cases reported in Los Angeles:

Date of diagnosis	AIDS cases reported in period
Jan–June '83	116
July–Dec '83	154
Jan–June '84	197
July–Dec '84	269
Jan–June '85	415
July–Dec '85	503
Jan–June '86	668
July–Dec '86	773
Jan–June '87	952
July–Dec '87	933
Jan–June '88	955
July–Dec '88	943
Jan–June '89	967

Reprinted by permission.

To get a feeling for the data, plot the points corresponding to the table. Number the horizontal axis 0, 1, 2, . . . to count the time intervals in the table, with 0 corresponding to the period January–June 1983, and so on, and the vertical axis to count the numbers of new diagnosed AIDS cases in Los Angeles during that period. You will see that not all these points appear to lie along an exponential curve.

To create a model for this situation, we will need to splice two functions together. Here's a start. Using your graphing utility to help you, draw the graph of

$$j(x) = 121e^{.272x}$$

on the same axes where you plotted the points from the table. Does the function serve as a reasonable model for your data?

For what x-values is j **not** a good model? For these x-values, a much simpler function will serve to model the data. Give such a function and the interval on the x-axis where it is appropriate. Now use both formulas to write a *single* piecewise function that could represent the entire period 1983–1989. Be sure to indicate which formula goes with which years.

What might account for the striking difference between the model for women and the model for gay men in Los Angeles? One explanation could be the effects of a significant educational effort in the gay community. (Even so, we should note that the disease was continuing to spread at the rate of approximately 1900 new cases per year—this equals 950 every six months, or a new case every 4 1/2 hours.) Since women have not been thought to be especially at risk, educational efforts have not been directed to them as a group. Many people are hopeful that education will be the key to slowing and, eventually, reversing the exponential spread of AIDS in all groups of the population.

THE LABORATORY REPORT

Discuss the properties of exponential functions, including domain, range, asymptotes, and the relationship between the graph and the choice of base.

Summarize your observations on the use of exponential functions to model the two sets of AIDS data. What were your reactions to the exponential growth predictions? What limitations did you see in these models? Give your final version of a function to model the Los Angeles data for the entire period 1983–1989.

Illustrate your report with appropriate graphs, both of abstract exponential functions and of the models used in the lab.

Homework 9.1: Growth Rates of Exponential Functions

The number e is used so often in mathematics that it is called the *natural* base for the exponential and logarithm functions. The usefulness of e is most apparent when we apply the techniques of calculus to the study of exponential functions. But you will be able to learn one property of e from its graph. You have seen that when you zoom in on the graph of a polynomial or rational function, it looks more and more like a straight line. The same is true for exponential functions. (Check it out!) The three functions 2^x, e^x, and 3^x have the same y-intercept. (What is it?) Which of the graphs is the steepest there? Which is the shallowest?

1. We wish to find the slope of each of these graphs at this common point. Zoom in on the y-intercept, or magnify several times, until the three graphs appear to be straight lines. Use your graphing utility to find two points close together on the graph of 2^x but on either side of the y-intercept, and use their coordinates to approximate the slope of the curve at the y-intercept. (In order to get accurate results, you should use all the decimal places available when you write the coordinates of the two points and calculate the slope.) Do the same for the slopes of the graphs of e^x and 3^x at their y-intercepts. When you finish, round off the values for the slopes to two decimal places.

2. What is your best guess for the slope of the graph of $g(x) = e^x$ at its y-intercept? What is the value of g at that same point?

3. Pick another point, not near the y-intercept, on the graph of $g(x)$. Write down the value of g, correct to two decimal places, at that point. Then, by magnifying, estimate the value of the slope of the graph of g at that point. (Magnify several times; use all the accuracy the graphing utility provides; round your final answer to two decimal places.) Compare that number to the number you wrote for the value of g.

   Well! Was that last result a fluke? Pick some other point on the graph and repeat the above procedure.

   What you are witnessing is the property of e that makes it unique among numbers:

The rate of growth (slope) of the function e^x at any point is equal to its value at that point.

   This is why, whenever we study a process of continuous exponential growth, the number e is involved.

   You might wonder whether the slopes of the other two functions at their y-intercept have any significance. (What do you expect? This is mathematics—everything has significance!) Using your calculator, write the natural logarithms for 2 and 3 ($\ln 2$ and $\ln 3$). Compare those two numbers with the slopes you already found for 2^x and 3^x, at the point where they cross the y-axis.

Homework 9.2: Limits to Growth

As you saw in Lab 9, an exponential function can model a real-world situation over only a limited period of time. After a while, we begin to run out of people, or space, or money, or whatever it is whose growth we are studying, and the growth has to slow down; that is, the *rate* of growth must decrease.

Notice how the *rate* of growth of an exponential function like e^x or $y = 16e^{.586x}$ is *itself* increasing over time: the function e^x **increases at an increasing rate**. Not only is it getting bigger, it's getting steeper as well.

1. Make a free-hand sketch of a graph to model a growth process in which there's a limit to the growth. Let A_0 represent the initial amount and suppose that the limit is given by some number A_{max}. (Assume A_{max} is greater than A_0.)

   Think a bit before drawing the graph. If there's a limit to the growth, that limit will begin to exert its influence on the growth rate as the value of A approaches the limit. The graph shouldn't get progressively steeper and then suddenly slam into the "ceiling."

2. Describe, qualitatively, what happens over time to the growth *rate* of $A(t)$. Refer to your sketch.

In Lab 9, you used the exponential-growth function $A(t) = ce^{kt}$ to model the number of deaths from AIDS in the population of young women, and you observed no slowing down of the growth. As time passes, however, and educational and preventive measures intensify, the rate of spread of the disease in that population is bound to abate. Let us suppose that it is extremely unlikely that more than 7000 women will ever die of the disease in a given year. A model that reflects this assumption is

$$A(t) = \frac{7000}{1 + 388e^{-.586t}}$$

3. Using the graphing utility, examine this function, which is called a **logistic growth model**. (**Caution:** enclose the entire denominator, as well as the exponent, in parentheses.) Does it have a shape similar to your free-hand sketch? (Try several different intervals.) How many deaths does it give for 1989 and 1990? Compare these values with the 1989 and 1990 values given by the straight exponential function $y = 16e^{.586x}$.

4. Compare the two functions over the long run. What values do they have after 20 years? after 50 years?

5. What is the approximate rate of growth, in deaths per year, of $A(t)$ when $t = 25$? What is the approximate rate of growth, in deaths per year, of the straight exponential function when $x = 25$? (Here you need to determine the *slopes* of the functions at 25, rather than their values at 25.)

6. Write a brief paragraph comparing and contrasting the logistic growth model and the exponential growth model, emphasizing the way in which their growth rates behave.

Homework 9.3: "Vamoose"

PERCENTAGE GROWTH AND DECLINE

A homework problem from Lab 1, "A Big Moosetake," contains data on automobile-moose collisions. The alarming increases in car-moose collisions in northern New England is due to growth in the moose population. Reforestation of land previously used for farming and a decrease in the deer population (deer carry a parasite deadly to moose) are contributing factors to the increased numbers of moose in the region.

1. Suppose the moose population were to grow at a rate of 10% each year. At the start of 1991, there were an estimated 4000 moose in New Hampshire. How many additional moose would there be one year later? What would be the moose population at the start of 1992? at the start of 1993?

       Instead of calculating the moose population year by year, you will write a formula that estimates the number of moose in a given year t. Let $t = 0$ correspond to the year 1991. The number of moose in subsequent years, according to a 10% yearly growth rate, can be modeled using an exponential function of the form

$$f(t) = c \cdot e^{kt}$$

2. Using two of the numbers you already have (moose populations at the start of 1991 and 1992, for instance), determine values for c and k. (Is k positive or negative?) $f(1)$ should coincide with the number of moose that you estimated for 1992 and $f(2)$ with the number estimated for 1993.

Approximately how many moose would you expect in New Hampshire in 2001 if this growth process were to continue unchanged over the next 10 years?

Suppose that, in the year 2001, developers discover New Hampshire and begin to build on some of the forest land and the deer population starts to rise. These changing conditions cause the moose population to **decrease** at a rate of 15% a year after the year 2001. Thus, after $t = 10$, the number of moose is governed by a new exponential relationship of the form

$$f(t) = c \cdot e^{kt}$$

for a different choice of c and k.

3. Use the number of moose in 2001 (from your first function) and the number in 2002 (from your hand calculations) to determine the new values for c and k, applicable after $t = 10$. You're going to have to be careful here: $t = 0$ must still correspond to 1991 if this formula is to be compatible with the previous one.

4. Sketch the graph of the piecewise function that models the changing number of moose in New Hampshire over the 30-year period from 1991 to 2021. Be sure to provide scales and to label the axes. Assuming that this rate of decrease continues, predict the year by which the moose population in this region will be practically nonexistent. (You'll have to decide for yourself what constitutes being practically nonexistent.)

CONSTANT RATES OF GROWTH AND DECLINE

In the previous example, we assumed that the moose population increased or decreased by a certain percentage each year. Suppose, instead, that the moose population in New Hampshire increases at the constant rate of 100 moose per year from 1991 to 2001. After 2001, the population begins to decrease at a constant rate of 150 moose per year.

5. Write a formula that models the number of moose in New Hampshire, with respect to time, for the years from 1991 until the moose have disappeared from the region. Your formula should be written as a piecewise linear function.

6. Sketch the graph of this function. Provide scales and labels.

According to this model, during what year will the moose disappear from New Hampshire?

Laboratory 10

PREPARATION

From 1943 to the late 1950s, the United States operated a secret scientific enterprise for developing materials for atomic weapons that covered more than 100 sites in 32 states and the Marshall Islands in the South Pacific. Since 1988, the Department of Energy has struggled to account for the astonishing array and volume of radioactive wastes that each site produced.

From South Carolina to Alaska, the program to clean up wastes left by the atomic weapons industry has become one of the Government's most technically difficult and expensive problems. The program is expected to cost $200 billion. . . .

Nearly a million cubic yards of dirt contaminated with thorium, uranium and radium, enough to fill much of Shea Stadium, are spread across abandoned processing plants in two sites near Buffalo and seven cities in New Jersey. More than two million cubic yards of spoils contaminated with radioactive elements and toxic chemicals are piled in five dumping grounds in the St. Louis area. In the West, enormous piles of radioactive sand produced by uranium processing mills are waiting to be sealed beneath tons of dirt and desert stones.[1]

"In the Trail of the Nuclear Arms Industry"
The New York Times, August 26, 1990.

Radioactive substances decay over time by emitting alpha and beta particles and gamma rays. The Nuclear Regulatory Commission (NRC) uses special computer software that holds information on hundreds of radioactive substances and will model various conditions at storage sites (such as ground water and soil types) in an effort to manage the containment of radioactive substances responsibly. Have you ever questioned how long radioactive materials would need to be cared for before they are safe?

The **half-life** of a radioactive substance is the time it takes for half of the substance to decay (into other substances, some of which are also radioactive). Let's assume that 10 grams of Uranium-232 is found at one of the sites. U-232 has a half-life of 72 years. Thus, after 72 years only 5 grams of U-232 would be left (the missing 5 grams having decayed into other substances or lost as released energy). After 2 half-lives (144 years) 2.5 grams of U-232 would remain. How much U-232 would remain after 3 half-lives (216 years)? 4 half-lives (288 years)?

On a piece of paper, sketch a graph of the relationship between the amount of U-232 and the number of half-lives (use the horizontal axis for the number of half-lives). You have just drawn an exponential decay curve for U-232. (In preparation for Lab 9, you drew a graph depicting the results of repeated doubling. Now you are examining the behavior of repeated halving.) Using your hand-drawn graph, estimate how much of the U-232 remains after $1\frac{1}{2}$ half-lives. (How many years is this?) The amount of U-232 present after t half-lives, $A(t)$, is of the form

$$A(t) = ce^{kt}$$

where c and k are both constants that need to be determined before you can use your graphing utility to analyze the decay curve.

Use the fact that $A(0) = 10$ to determine the value of c. Replace c with this value. Now use the information that $A(1) = 5$ to find the value of e^k. You can work on finding

---

[1]Copyright ©1990 by The New York Times Company. Reprinted by permission.

k in a subsequent homework assignment, but knowing c and e^k is enough to completely specify $A(t)$, since

$$A(t) = c(e^k)^t$$

is equivalent to the more usual form for $A(t)$ given previously.

Write down the completely determined decay function and bring it to the lab, along with your sketch of the graph.

LABORATORY 10: RADIOACTIVE DECAY

"Waste" Deep in the Big Muddy

DECAY CURVES USING HALF-LIVES

Graph the decay curve for U-232 using your graphing utility. Adjust the viewing window so that your graph will be in the same scale as the graph you drew by hand in the preparation section. Compare the hand-drawn graph to the graph on your viewing screen. Using your graphing utility, estimate to at least two decimal place accuracy the amount of U-232 remaining after $1\frac{1}{2}$ half-lives. (How do you know that you have achieved the desired accuracy?) Compare this estimate with the one you obtained from the hand-drawn graph.

How many years will it take before only 10% of the original amount of U-232 remains? (Be careful here; your function is in terms of half-lives of U-232. You must convert your answer from half-lives into years.)

When U-232 decays, it converts into other substances, which in turn decay into still other substances. A partial listing of the decay chain is given below.

Nuclide	Half-life
U-232	72 years
↓	
Th-228	1.91 years
↓	
Ra-224	3.62 days
↓	
Rn-220	55.6 seconds
↓	
Po-216	.146 seconds
↓	
Pb-212	10.6 hours

Eventually the chain stops with a nonradioactive, stable substance.

What if you started with 10 grams of thorium-228 (Th-228)? Determine the function relating the amount of Th-228 to time, as measured in thorium-228 half-lives. Sketch the graph. How does this function compare to the decay curve you had for 10 grams of U-232? How many years would you have to wait before only 10% of the original sample of Th-228 were left?

Repeat the work of the last paragraph, substituting Radium-224 (Ra-224) for Th-228. Do you have to work all the calculations from scratch or can you rely on patterns you have observed in the previous two examples?

You were asked to find the number of years it would take before only 10% of the original radioactive substance remains. To answer this question, you needed to convert half-lives into years, since the independent variable of the function, $A(t)$, measured time in terms of half-lives of the substance.

DECAY CURVES USING YEARS

At times, it may be more useful to work with a function that relates the amount of original material remaining to time measured in years, instead of in half-lives. (RADDECAY, a software package used at the NRC for training in methodology for radioactive waste management, is capable of producing decay curves with an independent variable of time measured in years, days, or even seconds for hundreds of radioactive substances. The user needs only to specify the substance, time unit, and time duration.)

Let's consider the substance U-232 first. Go back to the graph you sketched by hand. Underneath the 1, 2, 3, 4, and so forth on the horizontal axis (indicating the number of half-lives), write the equivalent measurements in years. You now have a graph of the relationship of amount of substance to number of years. In order to graph this function with the graphing utility, you need to determine an algebraic expression that describes it. This can be accomplished by the following two-step procedure.

Step 1

Convert the number of years into half-life measurement using

$$t(x) = x/72$$

where x is years and $t(x)$ is the equivalent time measurement in half-lives.

Step 2

After converting time in years to number of half-lives, use the function $A(t)$ to determine the amount of substance left.

The diagram below shows how to compose the functions t and A to create a new function, which we write as $A \circ t$ and which we call the composition of A and t.

start with	apply function t	apply function A
$x \quad \longrightarrow$	$t(x) \quad \longrightarrow$	$A(t(x)) = (A \circ t)(x)$

This new function $A \circ t$ has as its input the time measured in years and has as its output the amount of the substance remaining after x years.

Graph the function $A \circ t$. Use your graphing utility to determine how much of the substance remains after 5 years; after 200 years.

In a similar fashion construct the function relating the amount of Th-228 to time as measured in years (assume that you start with 10 grams of the substance). Graph the function and determine how much remains after 5 years.

Uranium-235 has a half-life of 703,800,000 years. Suppose that you start with 10 grams of Uranium-235. Sketch a graph of the decay curve with time measured in years. (You shouldn't need to use your graphing utility or even have to write out the decay function explicitly in order to sketch this graph.) Using information from this graph, comment on the implications for safe storage of this substance.

RATES OF DECAY

You've been studying the relationship between time elapsed, measured in years, and the amount of a specified radioactive substance remaining. The relationship was specified by the function $A \circ t$. Now look at the *rate* at which the substance decays, rather than the quantity of substance remaining. How fast (in grams per year, grams per second, grams per half-life, or whatever units are appropriate) is the substance decaying? Some important questions to consider are these:

- As time passes, does the rate of decay stay the same, speed up, or slow down?

- Can you obtain this information by looking at the decay curves?

Let's see how you might find the answers to these questions. Recall that, when you looked at how fast a *linear* function was changing, you saw that its rate of change is constant and is measured by the slope of its graph. Now you will apply this idea to the decay curves.

Start with 10 grams of one of the radioactive substances mentioned above. Graph the decay curve with time measured in half-lives. Estimate the rate at which the substance is decaying after 1, 2, and 3 half-lives using the following approach. To approximate the rate of decay at, say, $t = 1$, zoom in on the point $(1, 5)$ on the graph until what you see in your viewing screen looks straight. (You will probably need to zoom in four or more times depending on the zoom factor of your graphing utility.) Find the coodinates of two points on this line (write down all the digits your graphing utility provides) and calculate the slope. Repeat this procedure for $t = 2$ and $t = 3$. Be sure to state the units in which the slope is measured. Does the rate at which the substance decays appear to remain constant, speed up, or slow down with the passage of time? Explain.

THE LABORATORY REPORT

Organize and report on the material that has been presented in this lab. Include a sketch and explanation of the decay curve that results from starting with 10 grams of a radioactive substance. How does the decay curve change when the independent variable is years rather than half-lives? How does measuring time in half-lives simplify the mathematics involved in radioactive decay? Compare the length of time it takes for all but 10% of U-232, Th-228, and Ra-224 to decay. Discuss how the rate at which the substance decays depends on time. Comment on potential problems in storing radioactive substances. Back up this comment with the mathematical evidence you acquired in doing this lab.

Homework 10.1: Changing the Initial Amount of Substance

How is the shape of the decay curve changed by altering the amount of radioactive substance that you begin with? The decay curve that resulted from an initial 10 grams of U-232 was

$$A(t) = 10(1/2)^t$$

with the independent variable t measured in U-232 half-lives. (Recall that the half-life of U-232 was 72 years.)

1. Suppose you start with 20 grams of U-232 instead of 10. Write the new $A(t)$ formula and sketch its graph. How many years would it take for all but 10% of these 20 grams to decay?

2. Now repeat the exercise in the last paragraph but change the initial amount to 5 grams.

3. Generalize the results you have just noted. Suppose you started with A_0 grams of U-232. Write down an expression for $A(t)$ in terms of A_0.

4. Sketch the graph of $A(t)$ by referring to the graph in Problem 2 and relabeling the vertical axis.

5. Determine how many years it would take before only 10% of the original amount of U-232 remained.

Homework 10.2: Repeated Doubling versus Repeated Halving

1. In "AIDS," Lab 9, you worked with the function 2^x, whereas in Lab 10 you worked with $(\frac{1}{2})^x$. Show both **graphically** and **algebraically** that $(\frac{1}{2})^x$ is just a reflection of 2^x. Specify what type of reflection you see. (You may want to refer back to Lab 3.)

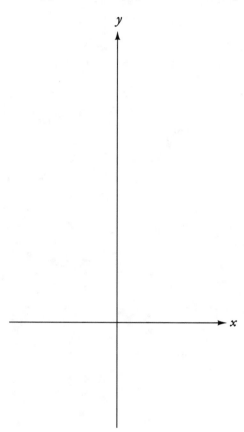

Homework 10.3: Exponential Decay

In Lab 10, you graphed the decay function

$$A(t) = 10(1/2)^t$$

where t was time in half-lives of the particular radioactive substance. This expression is algebraically equivalent to

$$A(t) = 10e^{kt} = 10(e^k)^t$$

for some value of k. You can find this value of k by solving $e^k = \frac{1}{2}$. There are several methods for finding k.

METHOD 1: APPROXIMATION FROM THE GRAPH OF e^x

1. Using your graphing utility, graph the function e^x. To get a rough estimate of k, use the TRACE feature of your graphing utility. Trace along the curve and observe the values that e^x assumes (y-values). Stop when you get as close as possible to $e^x = \frac{1}{2}$ and find the corresponding input value. What is your rough estimate for k? Is it positive or negative?

2. To get a more accurate estimate, overlay the constant function $f(x) = \frac{1}{2}$. Zoom in on the intersection of the constant function with e^x to get an estimate for k that is accurate to at least 4 decimal places. (How do you know that you have achieved the desired accuracy?) Write down your estimate.

METHOD 2: THE EXACT VALUE OF k

3. You need to determine the value of k for which $e^k = \frac{1}{2}$. In other words, to what power must you raise e to get $1/2$? Give your answer as a logarithm.

4. After you have written down the exact answer, use a calculator to estimate this value. What is your estimate of the exact solution? (Write down all the decimal places your calculator produces.)

5. Compare your estimate from Method 2 with your estimate from Method 1.

6. Substitute your estimate for k into $A(t) = 10e^{kt}$ and graph. Now overlay the graph of the function $10(\frac{1}{2})^t$. How do these graphs compare?

Laboratory 11

PREPARATION

The earthquake, which struck at 12:30 A.M. Thursday, was measured at Teheran University at 7.3 on the Richter scale, indicating a severe quake. It combined with more than 100 aftershocks to cause huge landslides that blocked roads and relief convoys, knocked out electricity and water supplies in many areas, destroyed crops and left many villages isolated and thousands of people buried in the rubble of their homes.[1]

The New York Times, June 23, 1990.

The map below shows the region described in the *New York Times* article. It pinpoints the earthquake's epicenter and outlines the regions hit hardest by the natural disaster.

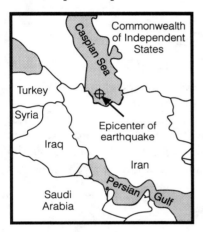

Earthquakes are dramatic events that attract a great deal of media attention. Coverage of earthquake disasters includes the death toll, reports of destruction, and estimates of economic impact, as well as specific information on the earthquake. Location of the earthquake's *epicenter* (the spot on the earth's surface directly above the region where ground movement first begins) along with its "size" as measured on the *Richter scale* are two pieces of information contained in nearly every report.

Have you ever wondered about the technical details contained in such news reports? For example, how did scientists determine that the epicenter for the 1990 Iranian earthquake was in the Caspian Sea? How much stronger in terms of ground movement was this quake than the 1989 earthquake in California, which measured 6.9 on the Richter scale?

Modern seismographs are located around the world (hundreds are in California alone) and record a number of different types of earth movements. Two types of motions that can be detected are P-waves and S-waves, both of which radiate out in rings from the epicenter. P-waves cause back and forth movements, whereas S-waves result in side to side motion. These two types of waves travel at different rates, the P-waves arriving first followed by the S-waves. Just as the time delay between thunder and lightning can be used to estimate

---

[1]Copyright ©1990 by The New York Times Company. Reprinted by permission.

distance from a storm, the time delay between the arrival of the P-waves and the S-waves can be used to determine the distance from the seismograph to the epicenter.

A seismograph, then, can measure both the strength of a quake and its distance from the seismographic station, but it cannot by itself determine the direction from which the P-waves and the S-waves are arriving. To know that, geologists need information from more than one station. Here's a geometric method for finding the epicenter of an earthquake.

The map on the following page has 5 seismographic locations. Assume that an earthquake generates the following data.

Location	Distance from epicenter
1	2 units
2	6
3	7
4	6
5	7

Using a compass, locate the epicenter of this hypothetical earthquake on the map. (If you do not have a compass, you can make one from a piece of string and a pencil. Tie the string to the pencil. From the pencil, mark off a length of string that is the desired distance from a particular location. Hold the mark at this location and pull the pencil away until the string is taut. Put the point of the pencil on the paper and draw an arc that results from keeping the string taut. Thus, every point on the arc is the desired distance from the location.) What was the minimum number of seismographic locations needed to pinpoint the epicenter of the earthquake?

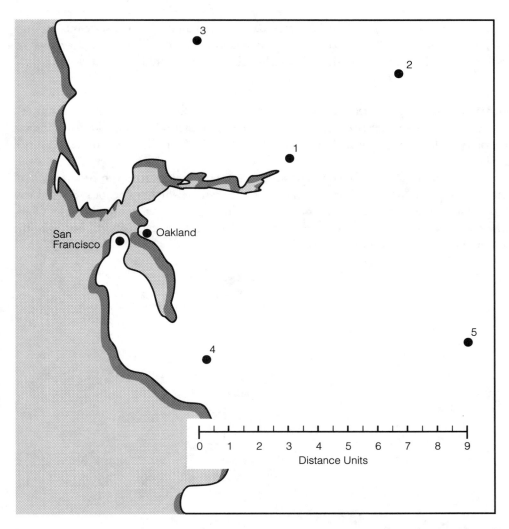

All earthquakes are compared to what is called a zero-level earthquake, that is, one that would produce readings of .001 millimeters on a seismograph located 100 miles from the epicenter. (Thus, an earthquake with seismograph readings of .02 is twenty times as strong as the zero-level earthquake, because $.02/.001 = 20$.) The following is a hypothetical list of seismographic readings. (Assume that the readings have been adjusted for distance, so that the numbers can be compared with one another.)

$$.003, \quad 10, \quad .02, \quad 1000, \quad .6, \quad 300$$

Try to plot these readings on a real number line. What difficulty do you encounter?

The magnitude of an earthquake, $M(x)$, as measured on the Richter scale is computed by taking the logarithm (base 10) of the ratio of the seismograph reading, x, to the zero-level reading, .001. You can express this relationship algebraically as follows,

$$M(x) = \log(x/.001)$$

where x is the seismograph readings in millimeters (adjusted for distance from the epicenter).

Compute the Richter scale magnitude of each hypothetical earthquake from the list of seismograph readings given above. Now plot these Richter scale magnitudes on a real number line.

When you get to the lab, compare worksheets with your partners to be sure you all centered the earthquake at the same location. How many seismographic locations did you need? Discuss the problems you encountered in plotting the "raw" seismograph readings. How did your conversion of these readings into their Richter scale equivalents make the data easier to represent on the number line?

LABORATORY 11: EARTHQUAKES

"Shake, Rattle, and Roll"

Because Richter magnitude is not well understood by the general public, reporters sometimes explain that an earthquake that measures 6 on the Richter scale is ten times as powerful as an earthquake that measures 5 on the Richter scale, and an earthquake that measures 7 on the Richter scale is 100 times as powerful as an earthquake that measures 5 on the Richter scale. Explain how the factors of 10 and 100 were determined. How much more powerful is an earthquake that measures 7 on the Richter scale than an earthquake that measures only 3 on the Richter scale?

Graph the function $M(x) = \log(x/.001)$ using your graphing utility. Theoretically, can this function ever be negative? For what values of x does this happen?

Although the portion of the graph where $M(x)$ is negative is important in terms of the abstract function, is this portion of the graph relevant to the function as representing the problem situation? Theoretically, how large can $M(x)$ get?

Now look at the data in the table listing major earthquakes of this century. What portion of the graph of $M(x)$ do you think actually represents the problem situation? (Give reasonable bounds for earthquake Richter scale readings and corresponding bounds for the seismograph readings.)

MAJOR EARTHQUAKES IN THIS CENTURY[1]

Year	Place	Richter scale	Deaths	Year	Place	Richter scale	Deaths
1989	California	6.9	62	1963	Yugoslavia	6.0	1,100
1988	Armenia	6.9	25,000	1962	Iran	7.1	12,230
1985	Mexico	8.1	9,500	1960	Chile	8.3	5,000
1983	Turkey	7.1	1,300	1960	Morocco	5.8	12,000
1982	Yemen	6.0	2,800	1957	Iran	7.1	2,000
1980	Italy	7.2	4,800	1957	Iran	7.4	2,500
1980	Algeria	7.3	4,500	1956	Afghanistan	7.7	2,000
1979	Colombia-Ecuador	7.9	800	1953	Turkey	7.2	1,200
				1950	India	8.7	1,530
1978	Iran	7.7	25,000	1949	Ecuador	6.8	6,000
1977	Romania	7.5	1,541	1948	Japan	7.3	5,131
1976	Turkey	7.9	4,000	1946	Japan	8.4	2,000
1976	Philippines	7.8	8,000	1939	Turkey	7.9	30,000
1976	China	7.8	200,000*	1939	Chile	8.3	28,000
1976	Italy	6.5	946	1935	India	7.5	30,000
1976	Guatemala	7.5	22,778	1934	India	8.4	10,700
1975	Turkey	6.8	2,312	1933	Japan	8.9	2,990
1974	Pakistan	6.3	5,200	1932	China	7.6	70,000
1972	Nicaragua	6.2	5,000	1927	China	8.3	200,000
1972	Iran	6.9	5,057	1923	Japan	8.3	100,000
1970	Peru	7.7	66,794	1920	China	8.6	100,000
1970	Turkey	7.4	1,086	1915	Italy	7.5	29,980
1968	Iran	7.4	12,000	1908	Italy	7.5	83,000
1966	Turkey	6.9	2,520	1906	Chile	8.6	20,000
1964	Alaska	8.4	131	1906	California	8.3	452

*Official estimate; may have been as many as 800,000
The New York Times, June 22, 1990.

While officials at Teheran University measured the Iranian earthquake to be 7.3 on the Richter scale, the United States Geological Survey at Golden, Colorado, reported a measurement of 7.7. Use your graph of $M(x)$ to determine how many times stronger the American estimate of the Iranian earthquake was than that indicated by the Teheran report.

The strongest aftershock of the Iranian earthquake measured 6.5 on the Richter scale. How many times stronger was the 1989 Californian earthquake than this aftershock?

In addition to the Richter scale magnitude, the table contains the estimated number of earthquake-related deaths. Does there appear to be any correlation between the number of deaths and the Richter magnitude?

Two articles appeared on the same page of *The Boston Globe* on August 16, 1991. "Nuclear weapon tested in Nevada" reported the detonation of a nuclear weapon 1600

---

[1] Copyright ©1990 The New York Times Company. Reprinted by permission.

feet underground. The test caused the ground to sway, registering 4.4 on the Richter scale. **"Quake put at 3.0 hits central Pa."** reported on a light earthquake with epicenter 10 miles from State College. The earthquake that measured 3.0 on the Richter scale was the strongest to hit this area since 1944. Please comment on these reports.

Earthquake measurement is only one example of logarithmic scale. Measuring the loudness of sound in decibels and the acidity of rain on the pH scale are two other examples. For Lab 5, you studied the Bordeaux Equation, in which the output Q was a logarithm (that is, an exponent), and needed to be put into the form e^Q to yield a meaningful number. In that lab, you saw bunched results (Q-values) become nicely spread out when they were converted to e^Q-values. In this lab, you saw input data (ground movement measurements) that were very spread out but that yielded meaningful numbers (Richter magnitudes) after logarithmic scaling.

THE LABORATORY REPORT

Describe how you used geometry to locate the epicenter of an earthquake and how you determined the minimum number of seismographic stations needed to pinpoint its location. Explain to someone not in this course how the Richter magnitude is determined from the seismograph reading and how it describes earthquake strength. Be sure to include reasons *why* we would want to use a logarithmic scale in this case. Support your reasons with specific examples. Show how to use Richter magnitudes to compare the strengths of two earthquakes. Illustrate your discussion with one or more graphs of the Richter function $M(x)$ and indicate the portion of the graph that gives reasonable bounds for Richter earthquake magnitudes.

Finally, report any other observations your group made. Did there appear to be a relationship between the magnitude of an earthquake and the number of resultant deaths? (Include specific examples and speculate as to other factors, besides Richter magnitude, that might contribute to the differences in death tolls.) What were your reactions to the two *Boston Globe* items from August 1991?

Homework 11.1: Explorations of ln x

1. Consider the function $f(x) = \ln x$. Sketch the graph of $\ln x$.

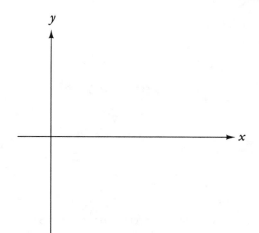

Use a viewing window with $-1 \leq x \leq 10$. (Decide for yourself how you want to adjust the vertical scale.) What is the domain of $f(x)$? For what values of x is $\ln x \leq 0$? $\ln x = 0$? $\ln x \geq 0$?

2. What happens to the value of $\ln x$ as you input values of x that are close to 0? Change your viewing screen to isolate the behavior of the function near zero. Try using a window with x between 0 and .01. (If your graphing utility allows, set y-values to be between -10 and -6.) Sketch this view of $\ln x$. To get a feel for how rapidly the natural log of x is falling as x values are chosen closer and closer to zero, fill in the following table.

x	$\ln x$
.01	
.001	
.0001	
.00001	

Try plotting these points by hand. What problem do you encounter?

3. Let's look at how this function behaves if we back away and look at the graph from a distance. Set the viewing window so that $0 \leq x \leq 100$. (You may have to adjust the *y*-values of the viewing window.) Does the graph appear to level off or keep rising? Repeat, using a viewing window with $0 \leq x \leq 1000$.

4. Find the value of *x* for which $f(x) = 6$. [You may want to overlay the graph of the constant function $g(x) = 6$ and then zoom in on the point of intersection between the two functions.] Find the value of *x* for which $f(x) = 7$ and for which $f(x) = 8$. How many *x*-units were there between $f(x) = 6$ and $f(x) = 7$? between $f(x) = 7$ and $f(x) = 8$?

5. Does $f(x)$ ever get bigger than 20? 100? (Can your graphing utility help you answer this last question?) Can you put a ceiling on how large the values for ln *x* can get? If so, what would the value of the ceiling be?

Homework 11.2: Algebraically Equivalent Functions

1. Graph $f(x) = e^{-\ln x}$. Clear your screen and graph $g(x) = \frac{1}{x}$. How are the graphs of $f(x)$ and $g(x)$ alike? (At this point you may want to overlay the two graphs to check your impression.)

2. What is different about these two functions? Give an algebraic explanation for what you determined about these two functions from their graphs.

3. Consider the function $f(x) = \ln x^2$. Specify the domain of $f(x)$. Discuss any symmetry inherent in this function. Graph $f(x)$.

4. Now consider the function $g(x) = 2 \ln x$. State the domain of $g(x)$. Graph $g(x)$. Is $f(x) = \ln x^2$ equivalent to $g(x)$? Explain why or why not.

5. Instead, let $f(x) = \ln x^3$ and $g(x) = 3 \ln x$. Is $f(x)$ equivalent to $g(x)$? Provide both graphical and algebraic support for your answer.

Homework 11.3: Alternate formula for $M(x)$

In Lab 11, the relationship between the seismograph readings x and Richter scale magnitude $M(x)$ was given by

$$M(x) = \log(x/.001)$$

Sometimes $M(x)$ is presented in the form

$$M(x) = \log(x) + 3$$

1. Graph the two expressions for $M(x)$.

2. How do the two graphs compare?

3. Show algebraically why this happens.

Homework 11.4: "I Hear the Train a-Comin' ..."

1. The human ear is sensitive to a wide range of sound intensities. Sound becomes audible after it reaches an intensity of 10^{-16} watt/cm^2 and becomes painful to the ear at about 10^{-4} watt/cm^2 (the threshold of pain). How many times greater is the sound intensity at the threshold of pain than the sound intensity for the faintest audible sound?

In Lab 11, you saw that measurements of earthquake strength given by seismograph readings, x, became more manageable when converted to a logarithmic scale, the Richter scale. Specifically, this relationship was given by the function

$$M(x) = \log(x/.001)$$

Similarly, converting sound intensity into a logarithmic scale (base 10) produces data that is more meaningful than the watt/cm^2 data. The sound intensity level $\beta(x)$, measured in decibels, is related to the sound intensity x, measured in watts/cm^2, by the formula

$$\beta(x) = 10\log(x/x_o)$$

where $x_o = 10^{-16}$ watt/cm^2 (a reference intensity corresponding to the faintest sound that the average person would be able to hear).

2. What would be the sound intensity level in decibels for the faintest sound that can be heard by the average person? What would be the sound intensity level for the threshold of pain?

3. The sound intensity level of a quiet whisper is about 20 decibels, whereas an ordinary conversation is a bit more than three times as loud, or about 65 decibels. Compare the sound intensities, measured in watts/cm^2, for the quiet whisper and ordinary conversation.

4. A difference in sound intensity level of one decibel is generally taken to be the smallest increase in volume that the average human ear can detect. How much of an increase in watts/cm² will be required to raise the sound level of a 20-decibel whisper by one decibel? How much of an increase will be required to raise the level of a 65-decibel conversation by one decibel?

5. The sound of a train passing over a trestle measures about 90 decibels. Suppose a particular train caused a noise of 91 decibels. Would the increase in watts/cm² corresponding to a change from 90 to 91 decibels be more or less than the increase required to go from 65 to 66 decibels?

6. Examine the shape of the graph of $\beta(x)$. As in "Earthquakes," you will need to use several different (reasonable) intervals in order to get a complete picture of a logarithmic curve. What happens to the graph as the sound intensity gets very close to zero? How does the basic shape of this graph compare to the graph of $M(x)$ sketched for Lab 11? Comment on the similarities between the formulas for $M(x)$ and $\beta(x)$.

Laboratory 12

PREPARATION

Before coming to the lab, you should become familiar with the graph of the sine function. You should know the meaning of the terms **amplitude**, **period**, and **phase shift**. Sketch the following graphs. Show scales on the axes.

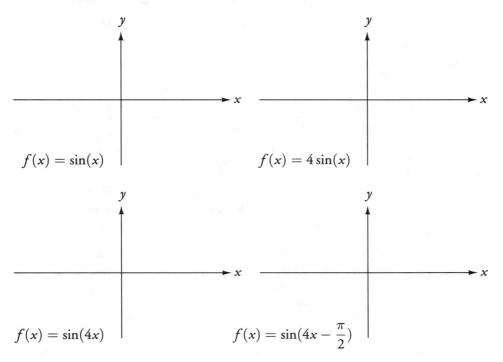

$f(x) = \sin(x)$

$f(x) = 4\sin(x)$

$f(x) = \sin(4x)$

$f(x) = \sin(4x - \dfrac{\pi}{2})$

The last one might be tricky. Be sure you have the correct phase shift.

If your graphing utility can't plot individual points, you should also prepare, on graph paper, two graphs of daylight data from the tables in the lab sheets: one for Boston and one for Fairbanks. To be able to see the annual pattern more clearly, plot two years' worth of data. **Bring the graphs with you to the lab.**

If the graphing utility does have point-plotting capability, plan ahead how to enter the data so that you can see a two-year curve.

LABORATORY 12: DAYLIGHT AND SAD

"You Are My Sunshine"

Over the course of a year, the length of the day—that is, the number of hours of daylight, calculated by subtracting the time of sunrise from the time of sunset—changes every day. Here is a table giving the length of day, rounded off to the nearest tenth of an hour, for Boston, latitude 42° N.

Date	Day	Daylight hrs	Date	Day	Daylight hrs
12/31	0	9.1	7/9	190	15.1
1/10	10	9.3	7/19	200	14.9
1/20	20	9.6	7/29	210	14.6
1/30	30	9.9	8/8	220	14.2
2/9	40	10.3	8/18	230	13.8
2/19	50	10.8	8/28	240	13.3
3/1	60	11.2	9/7	250	12.9
3/11	70	11.7	9/17	260	12.4
3/21	80	12.2	9/27	270	11.9
3/31	90	12.7	10/7	280	11.5
4/10	100	13.1	10/17	290	11.0
4/20	110	13.6	10/27	300	10.6
4/30	120	14.0	11/6	310	10.1
5/10	130	14.4	11/16	320	9.8
5/20	140	14.8	11/26	330	9.5
5/30	150	15.2	12/6	340	9.2
6/9	160	15.3	12/16	350	9.1
6/19	170	15.3	12/26	360	9.1
6/29	180	15.3			

Plot these points (day of year versus hours of daylight) on your graphing utility, if it has point-plotting capability. Otherwise, plot them on graph paper. Plot two years' worth of data (assume 365 days to the year).

The shape you see should look like a rough approximation of a sine wave. In fact, the graph can be approximated by a function of the form

$$f(x) = A \sin(Bx + C) + D$$

Our challenge here is to determine values for the constants A, B, C, and D.

First, we might look at A, the amplitude. We can find the amplitude of a sine function by taking half the difference between the highest point and the lowest point on the curve. Since we don't have data for every day of the year, and since the values have been rounded off, we don't know exactly what the highest and lowest values are or on which days they occur. What might be reasonable numbers to try for the highest and lowest values, given the data that we do have?

Next, let's try to determine B, which provides information on the period of the function. What is the formula for the period of any sine function? What is the period of this function? (After how many days does the length-of-day pattern repeat itself?) Knowing the period of this function and the general formula, you can solve for B.

Skip over C for the moment and consider D. The constant D tells how far up or down from the x-axis the graph is shifted. Compare this sine curve with a standard sine curve of the same amplitude and period. How far up has this one been shifted? How could you use the maximum and minimum values to find out?

Once you determine D, you have the horizontal axis of the sine wave (plotting the line $y = D$ would be helpful) and you can begin to guess at C. C provides information about the phase shift. What is the formula for the phase shift of any sine function? What value of C would cause a shift of one unit to the right? You need to estimate how far to the right or left a standard sine wave was shifted to end up in its current position. The horizontal axis $y = D$ will help. Then use your estimate of the shift and your value for B to solve for C.

Compare your value for C with those of your lab partners and others in the lab with you. You probably won't all have exactly the same number. Remember, the data are approximations and so were the points you plotted. If you used a graphing utility to plot the points, you should notice that few of the points went exactly where you tried to place them. This is because there are a limited number of pixels on the screen, and so the graphing utility does the best it can. Much of what you are doing in choosing the constants involves intelligent estimation. You are looking for a function that models the data fairly well. There is more than one such function, and no function will be a perfect match.

Try out your function; that is, plot $f(x) = A \sin(Bx + C) + D$. How well does it match the data? You may wish to adjust one or more of the constants to get a better fit. One difficulty inherent in the problem is that the given pattern is *not* a perfect sine wave, since the apparent path of the sun is actually one that is slightly elliptical rather than perfectly circular. This is a problem we are not going to attempt to fix in this lab.

The seasons affect everyone's moods to some degree, but some people are so strongly affected by the amount of daylight that they experience severe depression during that part of the year when the hours of daylight are shortest. Many magazine articles have been written about this condition, which has been termed seasonal affective disorder, or SAD. A typical person with SAD feels depressed for two or three months, sometime between the end of October and late February. She (women are affected more often than men) may experience a lack of energy and a craving for carbohydrates, and she may respond by oversleeping, overeating, and withdrawing from society. An estimated 6 to 8% of the population of New England suffers from full-blown SAD.

Unlike the traditional treatments for other forms of depression, an effective therapy for SAD has the patient sit in front of bright lights every morning. If we assume that one hour of light therapy is equivalent to an hour of natural daylight, approximately how many hours of light therapy might a person with SAD require on a day in early January if she wanted to make up for the "missing" hours of natural daylight (compared to March 21, the first day of spring)?

The graph that follows represents results from a study[1] of SAD patients and a group selected at random from the New York City telephone book, in which they were asked to specify the months in which they felt best or worst. Each point shows the proportion of people feeling at their best or worst in a particular month. "Feeling worst" is counted as a negative value.

If the seasonal mood fluctuations of SAD patients could be approximated by a sine function, would the function be in phase with the length-of-day function or out of phase? Explain.

Notice that the effect is reversed during spring and summer: SAD patients may, in fact, feel better than the average person. They are full of energy and usually lose the extra weight they put on during the winter.

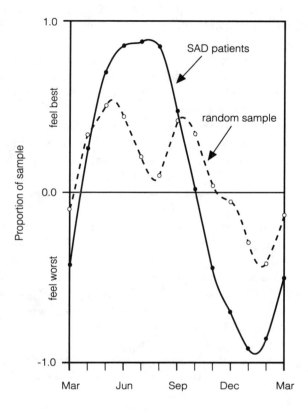

[1] Michael Terman (1988) On the Question of Mechanism in Phototherapy for Seasonal Affective Disorder: Considerations of Clinical Efficacy and Epidemiology, in *Journal of Biological Rhythms*, Vol.3, No.2, pages 155–172.

SAD appears to be even more prevalent farther north. Here are some data for Fairbanks, Alaska, latitude 65° N.

Day	Daylight hrs	Day	Daylight hrs
0	3.7	190	20.6
10	3.9	200	20.3
20	4.2	210	18.8
30	5.7	220	18.4
40	8.2	230	15.9
50	8.6	240	15.5
60	9.1	250	15.0
70	10.7	260	13.4
80	12.2	270	11.0
90	13.7	280	10.5
100	15.3	290	8.9
110	15.7	300	8.4
120	16.2	310	8.0
130	18.6	320	5.6
140	20.2	330	4.0
150	20.5	340	3.8
160	20.6	350	3.7
170	20.7	360	3.7
180	20.7		

When you plot these points, you will see that they do not seem to lie nearly so close to a smooth sine curve as Boston's points did. The Fairbanks data is far less precise; it was derived from the Boston data by adding some correction factors, which were themselves only rough approximations. Therefore, you must not take any single point too seriously. Nevertheless, it's possible to find a sine function that fits the data, taken as a whole, fairly well.

Writing such a function, $A \sin(Bx + C) + D$, won't involve much more work than you've already done. Think about what features of the graph remain substantially the same. Which letters (A, B, C, or D) control those features? What feature of the Fairbanks graph makes it different from the Boston graph? Which letter controls that feature? Calculate a value for that letter and try out your new function.

THE LABORATORY REPORT

Your lab report should explain how you decided on the value of each constant in the Boston length-of-day function and how you modified that function to write one for Fairbanks. Mention any difficulties you encountered in attempting to fit a sine curve to the data. Explain what information these functions might provide about seasonal affective disorder, including (among other things) the effect of latitude. Include any observations you made about the graph of seasonal mood fluctuations.

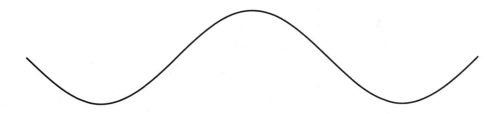

From 1948 until 1951 Japan practiced Daylight Savings Time. From April to September there was a summer time schedule of an additional hour of sunlight. This system was abolished because of the following reasons:

1.   The sun set too late. An additional meal was required because the day was so long.

2.   Longer hours for laborers.

3.   Lack of sleep.

More information is available on request.

—Japanese Embassy Information and Culture Center publication,
reprinted in The New York Times.

Homework 12.1: SAD and Latitude

The incidence and severity of seasonal affective disorder seem to depend upon latitude. *Scientific American* (January 1989) reports that 24% of the population of Tromsø, Norway (latitude 69° N.), may suffer from midwinter insomnia, another manifestation of SAD. Tromsø is so far north (more than 200 miles above the Arctic Circle) that the people there do not even see the sun between November 20 and January 20. (They do, however, enjoy 24 hours a day of sun during the summer.)

1. Draw a sketch of what the length-of-day function for Tromsø might look like. Is it a sine function?

2. In the Southern Hemisphere where the pattern of daylight is reversed, SAD reaches its peak during June and July. Draw the length-of-day graph for Wellington, New Zealand, or Puerto Montt, in south central Chile, both of which are approximately as far south of the equator as Boston is north of the equator. (In other words, their latitude is approximately 42° S.) Give a formula that could represent that graph. (Please note that this graph will be a modification of the graph for *Boston* that you studied in Lab 12.)

Homework 12.2: An Alternate Variety of SAD

Some people with seasonal depression are more prone to feel symptoms in spring and fall than in winter. Their depression appears to be triggered not by a shortage of light, but rather by changing amounts of it. Psychologists wishing to help such patients might want to know the rate at which the amount of daylight is changing at various times of the year.

1. Using the length-of-day function you wrote for Boston, examine the graph and find the season of the year during which the days are lengthening most rapidly. (How can you tell from the graph which time of year that is?) Zoom in on that section, magnifying several times, until the portion you see resembles a straight line. Use the trace feature to find the coordinates of two points on that line and calculate its slope. Use that information to determine how rapidly the days are lengthening at that time of year. Now convert your calculation into minutes per day, to the nearest whole minute. How many minutes per week is this? Would a person be likely to notice the difference from one week to the next?

2. Do the same steps for the part of the year when the days are shortening most rapidly. Compare the results with your previous ones.

3. Repeat the two sets of rate-of-change calculations, but this time use the length-of-day function for Fairbanks. Compare your results with those for Boston. Approximately how many times more rapidly are the days changing in Fairbanks?

4. Examine the region of the graph (either graph) where the greatest number of daylight hours occurs. What is the approximate rate of change, in minutes per day, for that region? Explain how you can read this information from the graph without having to do any calculations.

5. A psychologist studying this particular form of seasonal depression asks what you're working on. Explain, in a brief paragraph, the correlation between the two sine functions and this variety of SAD.

Homework 12.3: In the Bath

Engineers concerned with the public water supply need to know the daily water usage patterns for their community. A municipality typically has water storage tanks to provide a reserve for periods of high demand. When large quantities of water are being used, the amount of water stored goes down. During periods of light demand, the tanks refill from the municipal wells or reservoirs. Engineers monitor the water level in the storage tanks and can control the flow by opening or closing valves. They try to smooth out the flow so that a reserve is maintained.

The circular graph on the following page shows one week's readings from a device that measures the water level in a town's storage tank. The pen traces a path far from the center when the water in the tank is at a high level; the pen moves toward the center when the water is at a low level. Although there is some variation and some asymmetry, we can see a pattern here: the level is at its highest every morning and at its lowest every evening.

The circular graph was created when a disk of graph paper revolved slowly as a pen moved up and down. If we were to imagine the graph "unrolled," we could see time (in hours) on the horizontal axis and water level (in feet) on the vertical axis. The resulting curve would resemble a rough sine wave. Find a function that would approximate that sine wave. Even more than you did in Lab 12, you will need to take some liberties with the data. The amplitude is not constant from day to day; pick what looks to you to be a good average. The high and low points don't occur at the same time each day. In order to write a sine function, you need them to be equally spaced, so feel free to assign them to 8:00 or 9:00 or whatever seems right to you. Once you've made those leaps into imprecision, you'll be able to write a sine function just as you did in Lab 12. (We should note that such leaps have to be taken if we're to get anywhere. If we refuse to use anything but the exact numbers, the model we get to represent our reality will be just as large and unwieldy as the reality itself.)

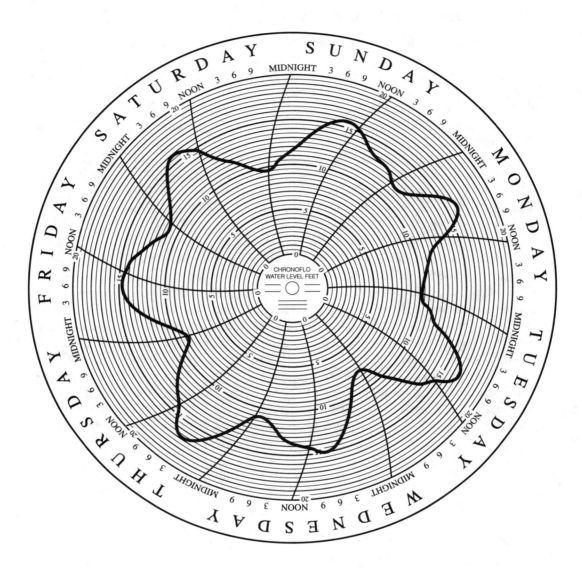

1. Write the formula that you came up with. Explain how you decided the value for each constant.

2. The water-level model (the idealized sine function you just wrote) also tells us something about water usage patterns in the town. Whenever more water is coming in, from wells or reservoirs, than people are using, the tank fills up. If people are using more water than is coming in, the tank empties.

Sketch the graph of your water-level function, labeling the axes. At what time of day does the water *usage* (not the water level) reach its maximum? At what time does it reach its minimum? Explain carefully how you determined your answers to these two questions.

Appendix

USING THE TI-81 GRAPHING CALCULATOR: A PRIMER

Basic Tutorial

- On, off, and contrast
- Basic calculation
- Correcting an error
- Graphing

Lab-by-lab specifics

Laboratory 1

- Graphing a linear function
- Plotting points

Homework 1.1

- True scaling

Laboratory 2

- Graphing a quadratic function
- Finding the coordinates of a point on the graph with $\boxed{\text{TRACE}}$

Laboratory 3

- Absolute value
- Use of parentheses
- Trigonomentric functions and the trig viewing screen

Laboratory 4

- Zooming in on a point to a desired accuracy
- Graphing piecewise functions

Homework 4.3

- Computing the least squares line

Programming the quadratic formula (Optional)

Laboratory 5

- No new techniques needed

Laboratory 6

- Zooming in using $\boxed{\text{BOX}}$

181

Laboratory 7

- Zooming out
- Changing the Zoom Factors

Homework 7.1

- Vertical asymptotes and holes

Laboratory 8

- No new techniques needed

Laboratory 9

- Exponential functions
- Scientific notation

Laboratory 10

- A word of caution

Laboratory 11

- Logarithmic functions
- Tracing beyond the viewing window

Laboratory 12

- Away dull trig tables!
- Problems inherent in the technology

BASIC TUTORIAL

As you proceed through this tutorial, note that specific keys you are to press appear in a box: $\boxed{\text{GRAPH}}$. Operations corresponding to the blue or white letterings above the keys are indicated in brackets: [ ]. To access upper key functions that are written in blue, press $\boxed{\text{2nd}}$ followed by the key. To access white upper key functions, press $\boxed{\text{ALPHA}}$ and then the key.

Work through the basic tutorial before starting the laboratories or homework from *Precalculus in Context.*

The second part of this primer introduces additional instructions relevant to the needs of individual laboratories and homework and should be worked through prior to, or in conjunction with, the particular lab or homework. This primer does not attempt to show you everything that can be done on the TI-81, so feel free to investigate other capabilities of your calculator by experimentation or with the aid of the TI-81 manual.

On, off, and contrast

Turn the calculator on by pressing $\boxed{\text{ON}}$.

You may need to adjust the contrast. Press $\boxed{\text{2nd}}$ followed by holding down the blue $\boxed{\uparrow}$ key to darken or the $\boxed{\downarrow}$ key to lighten.

To turn your calculator off press $\boxed{\text{2nd}}$ $\boxed{\text{[OFF]}}$. If you forget, the calculator will automatically turn off after a period of non-use.

Basic calculation

Press $\boxed{\text{CLEAR}}$ to start with a clear screen. You do **not** have to clear the screen after each computation.

Compute 3×4. After pressing $\boxed{3}$ $\boxed{\text{x}}$ $\boxed{4}$, press $\boxed{\text{ENTER}}$ in place of = on a standard calculator. Note that the original problem written as $3 * 4$ remains on the left side of the screen, and the answer appears to the right.

Compute $3 + 2 \times 6$ and $(3 + 2) \times 6$. In what order did the calculator perform the operations of addition and multiplication?

Compute 8^2 and 1.05^7. Press $\boxed{8}$ followed by $\boxed{x^2}$. You can also compute the square of 8 by pressing $\boxed{8}$ $\boxed{\wedge}$ $\boxed{2}$. Now try 1.05^7 using $\boxed{\wedge}$ $\boxed{7}$ to create the power.

Compute $\sqrt{2}$. Press $\boxed{\text{2nd}}$ $\boxed{[\sqrt{\ }]}$ (same key as x^2) followed by $\boxed{2}$.

Compute $\sqrt[3]{2}$. Press $\boxed{\text{MATH}}$ $\boxed{4}$ $\boxed{3}$ $\boxed{\text{ENTER}}$.

Compute $-2 + 4$. **Note:** Here you must be careful. On this calculator you must differentiate between the operation of subtraction (such as $3 - 2 = 1$) and the opposite of the positive number 2, namely, -2. Begin this problem by pressing $\boxed{(-)}$, then $\boxed{2}$ to create the number -2.

Correcting an error

Press 3 + + 2 ENTER . The following message will appear on your screen.

ERROR 06 SYNTAX
1:Goto Error
2:Quit

Press 1 and the cursor will locate the error. To erase one of the plus signs press DEL , then ENTER and the correct answer to 3 + 4 will appear.

Try 2 2nd [√] ENTER . This time press 2 to quit. Press CLEAR to clear the screen.

Press 3 + 4 ENTER . Suppose you really wanted −3 + 4. Press ↑ to go back to the previous command. Use ← to move the cursor over the 3. Press INS . (Note the cursor changes from a box to a line.) Now press (-) ENTER .

Graphing

Clear any previous work held in memory by pressing 2nd [RESET] . Warning: resetting will erase any stored programs.

The RESET menu will appear. Press ↓ to highlight 2: and then ENTER . You will probably have to adjust the contrast after resetting the calculator.

After resetting the calculator press RANGE (one of the keys directly under the viewing screen). Observe the settings for the standard viewing screen.

RANGE

```
Xmin=
Xmax=
Xscl=
Ymin=
Ymax=
Yscl=
Xres=
```

Press GRAPH and an empty viewing screen will appear. The scaling on the *x*- and *y*-axes go from −10 to 10 with tick marks one unit apart (because Xscl and Yscl = 1).

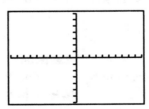

Get ready to try a few graphs.

Graph $y = x$. Press $\boxed{\text{Y=}}$ (the key to the left of $\boxed{\text{RANGE}}$). You should see the following display:

$$\boxed{\textbf{Y}=}$$

$$\boxed{\begin{array}{l} :Y_1= \\ :Y_2= \\ :Y_3= \\ :Y_4= \end{array}}$$

Press $\boxed{\text{X|T}}$ (for x) and then $\boxed{\text{GRAPH}}$. (A small black box in the upper right corner tells you that the calculator is still working on the problem. When the graph is completed, the black box will disappear. If you missed seeing the black box, watch for it on the next problem.)

Overlay the graph $y = x^2$. Press $\boxed{\text{Y=}}$ $\boxed{\downarrow}$ $\boxed{\text{X|T}}$ $\boxed{x^2}$ $\boxed{\text{GRAPH}}$.

Overlay the graph $y = x^3$. You need to enter x^3 to the right of $Y_3 =$.
Now, determine the points where the graphs $y = x$ and $y = x^3$ intersect.

First let's get a graph of these two functions alone. We can keep the function $y = x^2$ in the calculator's memory, but indicate to the calculator that we do not wish to see the graph of the parabola.

- Press $\boxed{\text{Y=}}$. Using the arrow keys, locate the blinking cursor over the equals sign next to Y_2 ($\boxed{\downarrow}$ followed by $\boxed{\leftarrow}$ will do the trick).

- Press $\boxed{\text{ENTER}}$. The highlighting over Y_2's equal sign has been removed.

- Press $\boxed{\text{GRAPH}}$ and note that the parabola no longer appears in the viewing window. (You can turn Y_2 back on by repeating the two previous steps.)

Where does the line $y = x$ intersect the cubic function $y = x^3$?

This is difficult to see in the standard viewing window. Press $\boxed{\text{ZOOM}}$ and the ZOOM menu will appear. A portion of this menu is shown below.

$$\boxed{\textbf{ZOOM}}$$

$$\boxed{\begin{array}{l} 1:Box \\ 2:Zoom \ In \\ 3:Zoom \ Out \\ 4:Set \ Factors \\ 5:Square \\ 6:Standard \\ 7:Trig \\ 8:Integer \end{array}}$$

- To zoom in on the center of the viewing screen press $\boxed{2}$ (there will be a blinking pixel at the origin). Press $\boxed{\text{ENTER}}$.

- To estimate the points of intersection, press $\boxed{\text{TRACE}}$. The cursor, a box with a blinking x through the diagonals, will appear on the graph of the line.

- Press $\boxed{\leftarrow}$ and watch the cursor box move down the line. The x- and y-coordinates of the cursor will appear at the bottom of the screen.

- Now press $\boxed{\rightarrow}$ to move along the line in the opposite direction.

- Press the $\boxed{\uparrow}$ or $\boxed{\downarrow}$ to jump back and forth between the line and the curve.

What are your estimates for the coordinates of the three points of intersection? How did the $\boxed{\text{ZOOM}}$ feature you selected affect the range settings?

- Press $\boxed{\text{RANGE}}$.

- Observe how $\boxed{\text{ZOOM}}$ affects the dimensions of the viewing window. Before zooming, the viewing window displayed an x-axis scaled from -10 to 10, a width of 20 units. After zooming, the width is approximately 5 units, a one-fourth reduction.

Here's another view of the graphs of $y = x$ and $y = x^3$.

- Reset the viewing window to match the one below.

RANGE

$$
\begin{aligned}
\text{Xmin} &= -3 \\
\text{Xmax} &= 5 \\
\text{Xscl} &= 1 \\
\text{Ymin} &= -6 \\
\text{Ymax} &= 6 \\
\text{Yscl} &= 2
\end{aligned}
$$

- Press $\boxed{\text{GRAPH}}$. The tick marks are one unit apart for the x-axis and two units apart for the y-axis.

To escape from any menu and get back to a clear screen, press $\boxed{\text{2nd}}$ $\boxed{\text{QUIT}}$. For example, press $\boxed{\text{Y=}}$. Now press $\boxed{\text{2nd}}$ $\boxed{\text{[QUIT]}}$.

That's it! You have completed the tutorial. Now practice and experiment with the calculator on your own until you begin to feel comfortable with these basic operations. The remainder of the primer will introduce new techniques when most appropriate for the work in *Precalculus in Context*.

LABORATORY 1

Before you get started, erase any functions you have stored.

- Press $\boxed{\text{Y=}}$. The blinking cursor will be on the line opposite Y_1.

- Press $\boxed{\text{CLEAR}}$ and the entire function will be erased.

Graphing a linear function

Graph $y = 2x + 3$. Set up the standard viewing window. (You may do this manually with [RANGE] or automatically with [ZOOM] [6] to select the standard window.)

- Press [Y=]. Enter the function $2x + 3$ to the right of $Y_1 =$.

- Now press [GRAPH].

Overlay the graph $y = 3x + 3$. How did changing the coefficient from 2 to 3 change the appearance of the line?

Plotting points

Although Laboratory 1 can be done by plotting various lines and then using [TRACE] to check the fit between the lines and the data in the table, the TI-81 will plot points. Thus, you can check how closely your guess at the relationship between Fahrenheit and Celsius matches the plot of the data. Keep in mind that the input variable is always x for the calculator and the output of the function is y. (What is the input variable for Lab 1? What is the output variable?)

Sample data

x	y
-2	40
-1	60
3	140
1	100

- Adjust the viewing window. The Xmin should be smaller than all the x-coordinates of the points you wish to plot and Xmax should be larger than all of the x-coordinates.

- Similarly, select appropriate settings for Ymin and Ymax.

- Decide on the spacing of the tick marks and set Xscl and Yscl. (What would be the disadvantage of setting Yscl=1? How many tick marks would appear between 40 and 140?)

Prior to entering the sample data you will need to clear any previously entered data from the calculator's memory.

- Press [2nd] [[STAT]].

- Using [→] highlight DATA. (Menu is shown below.)

```
DATA

1:Edit
2:ClrStat
3:xSort
4:ySort
```

- Press $\boxed{2}$ (ClrStat will appear on your screen) and then $\boxed{\text{ENTER}}$. The calculator will answer, "Done."

  Now you are ready to enter the sample data.

- Press $\boxed{\text{2nd}}$ $\boxed{\text{[STAT]}}$, highlight DATA, and press $\boxed{\text{ENTER}}$.

- Type in $\boxed{\text{(-)}}$ $\boxed{2}$ $\boxed{\text{ENTER}}$ $\boxed{4}$ $\boxed{0}$ $\boxed{\text{ENTER}}$.

- Enter the x-coordinates, then the y-coordinates of the remaining three points.

- To plot the points you have entered, press $\boxed{\text{2nd}}$ $\boxed{\text{[STAT]}}$. This time highlight DRAW.

- Press $\boxed{2}$ (to select a scatterplot) followed by $\boxed{\text{ENTER}}$.

Now graph $y = 18x + 85$. Note that the four data points have disappeared from the screen. To overlay the points on the same viewing screen as the graph of the line, repeat the keystrokes outlined in the previous two steps.

HOMEWORK 1.1: THE GRAPHING GAME

True scaling

The viewing screen on your calculator is a rectangle. Therefore, if you use the standard window, the tick marks on the y-axis will be closer together than the tick marks on the x-axis. For true scaling we want the distance between 0 and 1 on the x-axis to be the same as the distance between 0 and 1 on the y-axis.

- Set up the standard viewing window ($\boxed{\text{ZOOM}}$ $\boxed{6}$ will accomplish this) and graph the line $y = x$.

- Look at the spacing of the tick marks on the x- and y-axes.

- Press $\boxed{\text{ZOOM}}$ $\boxed{5}$ for Square, and observe the equal distance between tick marks on the two axes.

- Press $\boxed{\text{RANGE}}$ and note the changes for Xmin and Xmax.

- Now press $\boxed{\text{GRAPH}}$ to go back to the graph. The line should appear to be inclined at a 45° angle, cutting the 90° angle made by the intersection of the x- and y-axes in half.

LABORATORY 2

Graphing a quadratic function

It is important to experiment with various viewing windows to ensure that you have captured all the important features of the graph on your screen. Set the RANGE to the standard window settings.

Graph $y = 2x^2 - 2x + 12$. Note the empty viewing screen. Press RANGE and set up the following viewing screen:

RANGE

Xmin = 2
Xmax = 4
Xscl = 1
Ymin = 10
Ymax = 20
Yscl = 1
Xres = 1

Press GRAPH. Describe the shape of the graph that appears on the screen. Now change Xmin = 2 to Xmin = −2 and press GRAPH. You should see the familiar U-shape of a parabola.

Finding the coordinates of a point on the graph with TRACE

Suppose we need to find the coordinates at the bottom of the parabola $y = 2x^2 - 2x + 12$.

- Press TRACE and, using the → and ← keys, place the cursor at the vertex of the parabola.

- Read your first estimate of the coordinates of the turning point at the bottom of the graph.

- Press ZOOM 2 ENTER and again use TRACE, this time locating the turning point with greater accuracy.

LABORATORY 3

Absolute value

To graph $y = |x|$ press Y= 2nd [ABS] X|T GRAPH.

Use of parentheses

To graph $y = |x - 3|$ press Y= 2nd [ABS] (X|T − 3) GRAPH.

Note: To graph functions such as $\frac{x-3}{x^2-6}$ you will need to enclose the numerator and denominator in parentheses:

Press Y= (X|T − 3) ÷ (X|T ∧ 2 + 6) GRAPH.

Trigonometric functions and the trig viewing screen

Locate the $\boxed{\text{sin}}$, $\boxed{\text{cos}}$, and $\boxed{\text{tan}}$ keys.

- Press $\boxed{\text{Y=}}$ $\boxed{\text{sin}}$ $\boxed{\text{X|T}}$.

- To set an "ideal" viewing window automatically, press $\boxed{\text{ZOOM}}$ $\boxed{7}$ (for trig). You should see a nice sine wave.

- Press $\boxed{\text{RANGE}}$ and check the settings for this viewing window. (If you do not see a nice wavy curve, reset your calculator and start again.)

LABORATORY 4

Zooming in on a point to a desired accuracy

In the tutorial you were asked to find the three points of intersection between the graphs of $y = x$ and $y = x^3$. Those three points of intersection are $(-1, -1)$, $(0, 0)$ and $(1, 1)$.

Now let's use the calculator to estimate these points of intersection to two decimal-place accuracy.

- Set up the standard viewing window and graph $y = x$ and $y = x^3$.

- The three points of intersection between the line and the curve cannot be clearly seen. Press $\boxed{\text{ZOOM}}$ $\boxed{2}$. Move the cursor—four dots surrounding a blinking pixel—to the $(-1, -1)$ intersection point. Press $\boxed{\text{ENTER}}$. Again move the cursor to the leftmost point of intersection and press $\boxed{\text{ENTER}}$.

- Repeat a third and fourth time. Now $\boxed{\text{TRACE}}$ to the point of intersection and estimate the coordinates.

How can you be sure that the estimate for the x-coordinate is accurate to at least two decimal places? Press $\boxed{\text{RANGE}}$ and note that Xmin and Xmax agree to two decimals.

Graphing piecewise functions

Graph $y = \begin{cases} x - 3 & \text{for } x < 3 \\ -x + 3 & \text{for } x \le 3. \end{cases}$

Here you are asked to make a graph that consists of piecing two lines together.

You want to graph $y = x - 3$ for $x < 3$ and $y = -x + 3$ for $x \le 3$.

- Set your calculator for the standard viewing window.

- Press $\boxed{\text{Y=}}$ $\boxed{(}$ $\boxed{\text{X|T}}$ $\boxed{-}$ $\boxed{3}$ $\boxed{)}$

- Press $\boxed{(}$ $\boxed{\text{X|T}}$ $\boxed{\text{2nd}}$ $\boxed{\text{[TEST]}}$ $\boxed{5}$ $\boxed{3}$ $\boxed{)}$

- Keep going! Press $\boxed{+}$ $\boxed{(}$ $\boxed{\text{(-)}}$ $\boxed{\text{X|T}}$ $\boxed{+}$ $\boxed{3}$ $\boxed{)}$

- Press $($ $\boxed{\text{X|T}}$ $\boxed{\text{2nd}}$ $\boxed{\text{[TEST]}}$ $\boxed{4}$ $\boxed{3}$ $)$

- Press $\boxed{\text{ENTER}}$.

You'll need some explanation of the procedure you have just completed. The expression $(x \le 3)$ is assigned the value 1 when the inequality is true and the input variable, x, is less than 3; when the inequality is false, the expression $(x > 3) = 0$. Thus, for $x \le 3$ the function is equivalent to:

$$y = (x - 3)(1) + (-x + 3)(0) = x - 3.$$

And when $x > 3$ the following function will be graphed:

$$y = (x - 3)(0) + (-x + 3)(1) = -x + 3.$$

In the next example, you will learn how to enter a function as a variable.

Graph $y = \begin{cases} x^2 + 2x + 6 & \text{for } x \ge -2 \\ x + 8 & \text{for } x < -2. \end{cases}$

- Press $\boxed{\text{Y=}}$.

- Enter $x^2 + 2x + 6$ for Y_1 and $x + 8$ for Y_2.

- Graph these two functions in the standard viewing window.

You should see a line that intersects with a parabola. Now we want to draw a graph that consists of the line to the left of $x = -2$ and the parabola to the right of $x = -2$. Instead of retyping both functions, proceed as follows:

- Press $\boxed{\text{Y=}}$ $\boxed{\downarrow}$ to Y_3 $\boxed{\text{2nd}}$ $\boxed{\text{[Y-VARS]}}$ $\boxed{1}$

- Press $($ $\boxed{\text{X|T}}$ $\boxed{\text{2nd}}$ $\boxed{\text{[TEST]}}$ $\boxed{4}$ $\boxed{(\text{-})}$ $\boxed{2}$ $)$

- Press $\boxed{+}$ $\boxed{\text{2nd}}$ $\boxed{\text{[Y-VARS]}}$ $\boxed{2}$

- Press $($ $\boxed{\text{X|T}}$ $\boxed{\text{2nd}}$ $\boxed{\text{[TEST]}}$ $\boxed{5}$ $\boxed{(\text{-})}$ $\boxed{2}$ $)$

- Turn off the functions Y_1 and Y_2 (see instructions on page 185).

- Press $\boxed{\text{GRAPH}}$.

HOMEWORK 4.3: THE LEAST SQUARES LINE

Computing the least squares line

The sample data from the tutorial section of this primer lies exactly on a line, and, using algebra, you could write the equation for that line.

Plotted data often exhibits a roughly linear pattern. To find a line that best describes the pattern of the data points, start by entering the data below (see *Plotting Points* on page 187).

x	y
1.2	1.3
2.0	3.7
3.1	4.7
4.2	6.9

After entering the data press $\boxed{\text{2nd}}$ $\boxed{\text{[STAT]}}$ $\boxed{2}$ for **LinReg**, then press $\boxed{\text{ENTER}}$. The following will appear on your viewing screen.

LinReg
$a = \quad -.4242320819$
$b = \quad 1.742564603$
$r = \quad .9806596312$

The *y*-intercept is given by a and the slope by b. So the least squares line is $y = 1.742564603x - .4242320819$.

Graph this line and overlay a scatterplot of the data.

PROGRAMMING THE QUADRATIC FORMULA (OPTIONAL)

In Laboratory 4, you are asked to use the quadratic formula several times. The steps involved in applying the quadratic formula can be stored as a program in your calculator. Then the next time you need to use the quadratic formula, you can execute the program instead of re-entering all of the keystrokes.

Following is a program that will calculate the discriminant and roots of a quadratic function. The command lines have been numbered to the right of the command. Instructions on individual command lines follow the printout of the program. (It might be helpful to have the instructor available as you work through this exercise.)

Prgm1:QUADRATI	1
:Disp "INPUT A,B,C"	2
:Input A	3
:Input B	4
:Input C	5
:$B^2 - 4AC \rightarrow D$	6
:If $D < 0$	7
:Goto 1	8
:$(-B - \sqrt{D})/(2A) \rightarrow E$	9
:Disp "X="	10
:Disp E	11
:$(-B + \sqrt{D})/(2A) \rightarrow F$	12
:Disp "X="	13
:Disp F	14
:Lbl 1	15
:Disp "D="	16
:Disp D	17
:END	18

Before you begin: Trouble shooting

If you wind up in a menu that you had not intended to be in, press 2nd QUIT to return to the home screen. You can press PRGM to get back into the program you are writing.

If you need to insert a line in the program, move the cursor to the end of a line and press INS ENTER to insert a blank line below; or move the cursor under the colon (:) and press INS ENTER to insert a line above.

You can delete a line in your program with DEL.

Don't be afraid to turn to the section on programming in your TI-81 manual for additional help.

Word of warning: Do not reset your calculator or you will lose your program.

To program the quadratic formula

Press PRGM. Store the quadratic formula under Program 1: use → to highlight EDIT, then press 1.

Line 1. Name your program.

To the right of Prgm1: there will be a blinking box with an A inside. This means that the $\boxed{\text{ALPHA}}$ key has been activated.

Type in the name of your program, QUADRATI. (The letters appear in alphabetical order on your calculator with $\boxed{[A]}$ the white upper function of the $\boxed{\text{MATH}}$ key, and $\boxed{[Q]}$ the upper function of the $\boxed{9}$ key.)

Line 2. When you run this program, you will want to be prompted to input the values of the coefficients of the quadratic function, $Ax^2 + Bx + C$. Create the prompts as follows:

■ To access the Disp command, press $\boxed{\text{PRGM}}$ $\boxed{\rightarrow}$ to highlight I/O, and then press $\boxed{1}$.

■ To type in the remainder of line 2 press $\boxed{\text{ALPHA}}$ prior to each of the following: $\boxed{["]}$ $\boxed{[I]}$ $\boxed{[N]}$ $\boxed{[P]}$ $\boxed{[U]}$ $\boxed{[T]}$.

■ To get the space between INPUT and A press $\boxed{\text{ALPHA}}$ $\boxed{0}$. Note that the entire command does not fit on a single line and will wrap around onto the next line. (After you have entered all the characters for any command line, press $\boxed{\text{ENTER}}$.)

Lines 3–5. When the values of the coefficients are entered, they will be stored in memory locations A, B, and C.

To access the Input command press $\boxed{\text{PRGM}}$, use $\boxed{\rightarrow}$ to select I/O, $\boxed{2}$.

Line 6. The value of the discriminant is computed.

Enter the expression $\boxed{\text{ALPHA}}$ $\boxed{[B]}$ $\boxed{x^2}$ $\boxed{-}$ $\boxed{4}$ $\boxed{\text{ALPHA}}$ $\boxed{[A]}$ $\boxed{\text{ALPHA}}$ $\boxed{[C]}$ $\boxed{\text{STO}}$ $\boxed{[D]}$. (Note that after you press STO, the ALPHA key will activate.)

Lines 7 and 8. If the discriminant is negative there are no real roots, so you only want the discriminant to be computed and displayed.

When D is negative (line 7), the Goto 1 command (line 8) will be executed, causing the program to jump down to Lbl 1 (line 15) and to proceed executing command lines 16 and 17.

To access **If**, press $\boxed{\text{PRGM}}$ $\boxed{3}$. To get $\geq$, press $\boxed{\text{2nd}}$ $\boxed{\text{TEST}}$ $\boxed{5}$.

To access **Goto**, press $\boxed{\text{PRGM}}$ $\boxed{2}$.

Lines 9–14. If the discriminant is positive or zero, you want the program to compute and display the two roots (roots are equal if $D = 0$).

In line 9, remember to use $\boxed{(-)}$ for the opposite of B and the subtraction key $\boxed{(-)}$ between B and $\sqrt{D}$.

In line 10, to get the equals sign, press $\boxed{\text{2nd}}$ $\boxed{[\text{TEST}]}$ $\boxed{1}$.

Lines 15–17. Lbl 1 marks the place where the program is resumed if the discriminant is negative.

To access **Lbl**, press $\boxed{\text{PRGM}}$ $\boxed{1}$.

Line 18. Mark the end of the program.

To access **END**, press $\boxed{\text{PRGM}}$ $\boxed{7}$.

Press $\boxed{\text{2nd}}$ $\boxed{\text{QUIT}}$ to leave the program and return to the home screen.

To run the program, press PRGM. Note that EXEC and 1: are highlighted. All you have to do is press ENTER twice.

Using the quadratic formula program

Now let's find the roots of $2x^2 + 5x - 3$. Run the program you have just entered. A prompt appears on the screen.

INPUT A,B,C

?	press 2 ENTER
?	press 5 ENTER
?	press (-) 3 ENTER
X=	−3
X=	.5
D=	49

Debugging the program

If you have made errors, you are likely to get an error message when you try to execute the program. For example, if in computing the root on line 9 you had used the subtraction key for −B, the following screen will appear:

ERROR 06 SYNTAX
1:Goto Error
2:Quit

If you press 1, you will be taken back into the program and the cursor will locate the error. Correct the error and try running the program again.

LABORATORY 5

No additional instructions on the TI-81 calculator are needed for Lab 5 or the associated homework.

LABORATORY 6

Zooming in using BOX

In the Laboratory 2 section on pages 188–189, you were asked to locate the turning point on the graph of $y = 2x^2 - 2x + 12$ using ZOOM 2. Now we are going to experiment using BOX from the ZOOM menu.

Graph $y = 2x^2 - 2x + 12$. Make sure you have chosen a viewing window that clearly shows the U-shape of the parabola.

■ Press ZOOM 1 for Box. A blinking pixel will appear in the middle of your viewing window.

■ You are going to draw a small box around the turning point of the graph by locating the two diagonal corners of the box. Using the arrow keys, move the blinking pixel to the left and just a little above the bottom of the U. Press ENTER .

■ Now move the blinking pixel to the right and below the bottom of the U. A box should appear on your screen containing the turning point of the parabola (the local minimum). Press ENTER . The box becomes your new RANGE .

■ Repeat the above process starting with ZOOM 1 .

■ Now use TRACE to accurately estimate the local minimum of the graph.

Hint: If the box you draw is much wider than it is high, it can help you pinpoint the coordinates of the low point of the parabola.

LABORATORY 7

Zooming out

Graph $y = \frac{x^2-1}{x-3}$ in the standard viewing window.

What will happen to the appearance of the graph as we "back away" by increasing the width and length of the viewing window (by a factor of 4)?

■ Turn off the tick markings by setting Xscl and Yscl to 0.

■ Press ZOOM 3 for Zoom Out, ENTER .

■ Press RANGE and observe how the settings have changed.

■ Zoom out a second time. The graph should look like a line except for a small blip slightly to the right of the origin. (You will be able to see this better if you use the ← to move the cursor away from the center of the viewing window.)

Changing the Zoom Factors

■ Change the RANGE setting back to the standard and redraw the graph $y = \frac{x^2-1}{x-3}$. The default setting is for both the x- and y-interval to widen by a factor of 4 when you zoom out.

■ Let's change this setting. Suppose that as you zoom out you'd like to widen the x-interval by a factor of 4 but the y-interval only by a factor of 2. Turn off the tick marks by setting Xscl and Yscl to 0.

■ Press ZOOM 4 to select Set Factors.

ZOOM FACTORS

Xfact=4

Yfact=4

■ Locate the blinking cursor over the 4 opposite YFact and change the 4 to 2.

■ Press GRAPH ZOOM 3 ENTER .

■ Press ENTER a second time.

How did the change in the Zoom Factors affect the appearance of the graph as you zoomed out? Return the factors to the default settings.

HOMEWORK 7.1: BLACK HOLES AND VERTICAL ASYMPTOTES

Vertical asymptotes

Graph $y = \dfrac{x^2 - 1}{x - 3}$ in the standard viewing window.

Remember to enclose both the numerator and denominator in parentheses.
The vertical asympote at $x = 3$ has been drawn in for you. (This is not part of the actual graph of the function.) In this viewing window only the portion of the graph that lies to the left of $x = 3$ appears. Using RANGE , change Ymax to 20 and re-graph the function. You now have a more complete picture of how the function behaves in the vicinity of $x = 3$.

Holes

Graph $y = \dfrac{x^2 - 4}{x - 2}$.

Start with the standard viewing window and graph the function above.
This time there is no vertical asymptote. Instead there is simply a hole in the graph at $x = 2$. You need to know that this hole is present without relying on the calculator because, depending on your viewing window, you will not always be able to see the hole.
Press RANGE , set Xmax=15 and again graph the function. Can you still see the hole in the graph at $x = 2$?

LABORATORY 8

No additional instructions on the TI-81 are needed for Lab 8 or the associated homework.

LABORATORY 9

Exponential functions

Graph $y = 4^x$. Set the RANGE to the standard viewing screen.
Press Y= 4 ∧ X|T GRAPH . The base of this exponential function is 4. What happens when you try to use the TI-81 to graph an exponential function with a negative base?

Graph $y = (-4)^x$. Press Y= . Clear the function 4^x.

Enter $\boxed{(}$ $\boxed{(\text{-})}$ $\boxed{4}$ $\boxed{)}$ $\boxed{\wedge}$ $\boxed{\text{X|T}}$ $\boxed{\text{GRAPH}}$.
No, you are not going blind. Your screen is truly blank. (You might want to think about a possible explanation.)

The exponential functions with base 10 and e are so common that e^x and 10^x are builtin functions of the TI-81. Erase the previous function.

Graph $y = e^x$ **and** $y = 10^x$.

- Press $\boxed{\text{Y=}}$ $\boxed{\text{2nd}}$ $\boxed{[e^x]}$ (the LN key) $\boxed{\text{X|T}}$ $\boxed{\text{GRAPH}}$.
- Press $\boxed{\text{Y=}}$ $\boxed{\text{2nd}}$ $\boxed{[10^x]}$ (the LOG key) $\boxed{\text{X|T}}$ $\boxed{\text{GRAPH}}$.

Note: You must take care when interpreting the calculator-produced graphs of exponential functions. In places the graph is so steep that you might think that it has a vertical asymptote. In other places the function is so close to zero that the graph in your viewing window merges with the x-axis, even though the function is never zero. Let's check it out. Go back to the graph of $y = e^x$ in the standard viewing window. Look at the portion of the graph that lies to the left of $x = -1$. Here the graph of e^x appears to melt into the x-axis.
Now zoom in on the graph at this point: press $\boxed{\text{ZOOM}}$ $\boxed{2}$, use $\boxed{\leftarrow}$ to move the cursor over to the spot where the graph merges with the x-axis, and press $\boxed{\text{ENTER}}$.
Does the graph still appear to melt into the x-axis at about $x = -1$? With $\boxed{\text{ZOOM}}$ $\boxed{6}$ go back to the standard range settings. Does it look to you like the graph becomes vertical?
Re-graph the function using the following range settings:

$$\text{RANGE}$$

$$\text{Xmin} = -2$$
$$\text{Xmax} = 8$$
$$\text{Xscl} = 1$$
$$\text{Ymin} = 0$$
$$\text{Ymax} = 100$$
$$\text{Yscl} = 0$$
$$\text{Xres} = 1$$

In this viewing window is it clear that the graph does not become a vertical line?

Scientific notation

Scientific notation is extremely useful for representing numbers that are large in magnitude or very close to zero. For example, $3,230,000 = 3.23 \times 10^6$ and $.0000000056 = 5.6 \times 10^{-9}$. The calculator uses an abbrevation for scientific notation. For the first example, the calculator will display 3.23E6, the E6 signifying the $\times 10^6$ part. The second example will appear as 5.6E-9. Below are two contexts where numbers get displayed in scientific notation.

Compute 4^{20}. The answer reads something like 1.099511628E12, which represents the number 1,099,511,628,000.

Graph $y = 4^x$ in the standard viewing window.

- Zoom in on the y-intercept of the graph with $\boxed{\text{ZOOM}}\,\boxed{2}$.

- Use $\boxed{\uparrow}$ to pinpoint the y-intercept, then press $\boxed{\text{ENTER}}$.

- Again move the cursor over to the y-intercept and press $\boxed{\text{ENTER}}$ three more times.

- Move the cursor over to the y-intercept and read the x-coordinate. You should see something like $x = -4.112E - 4$. ($-.0004112$ is the closest pixel location to $x = 0$ that you can get using this viewing window.)

LABORATORY 10

A word of caution

In graphing $y = e^{(-x)}$, remember to use the $\boxed{(\text{-})}$ key to indicate to the calculator that you want the **opposite** of x for the exponent.

LABORATORY 11

Logarithmic functions

The logarithmic functions with base 10, $\log x$, and base e, $\ln x$, are builtin functions on the TI-81. Logarithmic functions of other bases can be graphed by dividing these functions by the appropriate scaling factor. As is the case with exponential functions, it is difficult to get a comprehensive picture of logarithmic functions from a single viewing window.

Graph $y = \log x$ in the standard viewing window. Press $\boxed{\text{Y=}}\,\boxed{\text{LOG}}\,\boxed{\text{X|T}}$. Now adjust the range settings to match the ones below.

RANGE

Xmin $=$ 0
Xmax $=$ 3
Xscl $=$ 1
Ymin $=$ -2
Ymax $=$ 1
Yscl $=$ 1
Xres $=$ 1

The curve descends so steeply near zero that the graph quickly appears to merge with the y-axis.

Tracing beyond the viewing window

- Now change Xmax to 100, Xscl to 0, and Ymax to 3. As the x-values increase, the curve becomes so flat that it is difficult to distinguish it from a horizontal line.

- Press $\boxed{\text{TRACE}}$ $\boxed{\rightarrow}$. Watch what happens to the y-coordinates as x gets bigger than 100.

- Keep pressing $\boxed{\rightarrow}$, and the $\boxed{\text{RANGE}}$ settings will automatically change to let you trace along the portion of the graph that lies outside the viewing window.

- Watch how gradually the y-values are changing with x-values larger than 150. Do you think the function will surpass 2.5? (Keep pressing the $\boxed{\rightarrow}$ key to find out.)

LABORATORY 12

Away dull trig tables!

Compute sin 45°. Press $\boxed{\text{SIN}}$ $\boxed{4}$ $\boxed{5}$ $\boxed{\text{MATH}}$ $\boxed{6}$ $\boxed{\text{ENTER}}$.

Compute cos 37° **and** tan 25°.

- Press $\boxed{\text{COS}}$ $\boxed{3}$ $\boxed{7}$ $\boxed{\text{MATH}}$ $\boxed{6}$ $\boxed{\text{ENTER}}$.
- Press $\boxed{\text{TAN}}$ $\boxed{2}$ $\boxed{5}$ $\boxed{\text{MATH}}$ $\boxed{6}$ $\boxed{\text{ENTER}}$.

Problems inherent in the technology (Don't believe everything you see!)

Set up the trig viewing screen with $\boxed{\text{ZOOM}}$ $\boxed{7}$.

Graph $y = \sin x$.

- Turn off the tick marks by setting Xscl and Yscl to 0.
- Press $\boxed{\text{GRAPH}}$.
- Change the Zoom factors: $\boxed{\text{ZOOM}}$ $\boxed{4}$. Set XFact = 10 and YFact = 1.

Get ready for some fun. If we increase the x-interval we should see more and more cycles of the sine wave. There are two complete cycles of the sine wave in the trig viewing window. How many should there be if we widen this interval 10 times?

- Press $\boxed{\text{ZOOM}}$ $\boxed{3}$ $\boxed{\text{ENTER}}$. Does the calculator's graph match your prediction?

If we widen the interval by 10 again, how many complete cycles of the sine wave should we see? Try it by pressing $\boxed{\text{ENTER}}$ again.

Press $\boxed{\text{ENTER}}$ 6 or more times. What kinds of patterns do you see on the screen? How do these patterns differ from what the graph of the sine function should look like in these intervals?